F-35 Block Buy

An Assessment of Potential Savings

James D. Powers, Guy Weichenberg, Abby Doll, Thomas Goughnour,
Thomas Light, Mark A. Lorell, Ellen M. Pint, Soumen Saha,
Fred Timson, Thomas Whitmore, Robert A. Guffey

Prepared for the United States Air Force
Approved for public release; distribution unlimited

For more information on this publication, visit www.rand.org/t/RR2063

Library of Congress Cataloging-in-Publication Data is available for this publication.

ISBN: 978-0-8330-9835-1

Published by the RAND Corporation, Santa Monica, Calif.

Cover: U.S. Air Force photo/Senior Airman Christine Groening.

Preface

The F-35 Lightning II is the most expensive acquisition program in the U.S. Department of Defense. It is intended to replace several fighter and attack aircraft for the U.S. Air Force, Navy, and Marine Corps, as well as those from a number of partner allied nations. The U.S. military services and partner nations are keenly interested in ways to reduce the cost of the program. The F-35 Joint Program Office asked RAND Project AIR FORCE (PAF) to analyze what savings might accrue to the program if three upcoming lots of aircraft were to be procured under a single block buy (BB) contract as opposed to multiple annual contracts. Similar to multiyear procurement contracting, BB contracting should provide the prime contractors and their suppliers the incentive and ability to leverage quantity and schedule certainty and economies of scale to generate savings that would not be available under annual single-lot contracting.

This document presents PAF's assessment of cost savings available to the F-35 program through a BB contract for three lots of F-35 aircraft scheduled to be contracted for fiscal years 2018, 2019, and 2020. As such, it is a key component of a business case analysis on whether the U.S. and partner governments should pursue a BB contract. Other considerations necessary for a full business case analysis, such as an assessment of risks associated with this contracting approach, are beyond the scope of this research. Related documents to this are:

- RR-2063/1-AF (Appendix B), which contains a set of detailed case studies documenting key aspects of previous multiyear procurement and BB contracts for other weapon system programs. It supports the content of Chapters Three, Four, and Six, along with Appendix A of this document.

- An appendix of supporting information, not available to the general public.
- RR-1706-AF, which presents a subset of the results from this document; specifically, an assessment of savings using an annual contracting baseline provided by the F-35 Joint Program Office. That summary document was intended for officials of the U.S. and partner governments interested in seeing a portion of PAF's results that may be relevant to their near-term decisionmaking.

The research reported here was sponsored by Lt Gen Christopher Bogdan, Program Executive Officer for the F-35 Lightning II Joint Program Office, and was conducted within the Resource Management Program of PAF. This document should be of relevance to those involved in the F-35 program and to those interested in methodologies for assessing cost savings in BB and multiyear procurement contracts.

RAND Project AIR FORCE

RAND Project AIR FORCE (PAF), a division of the RAND Corporation, is the U.S. Air Force's federally funded research and development center for studies and analyses. PAF provides the Air Force with independent analyses of policy alternatives affecting the development, employment, combat readiness, and support of current and future air, space, and cyber forces. Research is conducted in four programs: Force Modernization and Employment; Manpower, Personnel, and Training; Resource Management; and Strategy and Doctrine. The research reported here was prepared under contract FA7014-16-D-1000.

Additional information about PAF is available on our website: http://www.rand.org/paf/

This report documents work originally shared with the U.S. Air Force on July 1, 2016. The draft report, issued on September 30, 2016, was reviewed by formal peer reviewers and U.S. Air Force subject-matter experts.

Note from the F-35 Joint Program Office

The RAND F-35 block buy study and report was a key component to the successful implementation of the F-35 Joint Program Office strategy for production Lots 12-14. Since this work was conducted in the 2015–2016 time frame, some differences have arisen in the actual execution of the F-35 Lots 12-14 approach relative to the constructs analyzed in this report. The U.S. military services and partner nations are pursuing a contract strategy similar to the Hybrid 2 construct presented in Chapter Five. However, they are doing so with fewer aircraft than assumed in this report—442 aircraft over the Lots 12-14 period instead of 471 aircraft. In addition, to preserve congressional annual discretion, the U.S. military services are continuing to procure on an annual basis but are procuring material and equipment in economic order quantities for fiscal years 2019 and 2020. Lastly, there is no industry economic order quantity investment component in the F-35 current, as-implemented, approach. In spite of these differences, the analysis presented in this document supported the Lots 12-14 business case for cost savings and facilitated decisions by the Joint Strike Fighter Executive Steering Board, the Department of Defense budget process, and Congress.

Contents

Figures and Tables

Figures

Tables

Summary

The F-35 acquisition program is the most expensive in U.S. Department of Defense history, approaching some $400 billion at its completion. While the per-lot cost of the aircraft has been declining over time, the U.S. military services and participating partner nations are keenly interested in ways to further lower the program's cost. One proposal is to procure three upcoming production lots under a single block buy (BB) contract, rather than in multiple annual contracts.[1] A BB contract saves money by providing prime contractors and their suppliers the incentive and ability to leverage quantity and schedule certainty and economies of scale, thus generating savings that would not be available under three annual single-lot contracts. However, BB contracts are not without risks, which may include the availability of funding, configuration changes, and aircraft quantity reductions. To inform decisions about whether to pursue a BB contract for the upcoming production lots, the F-35 Joint Program Office (JPO) asked the RAND Corporation's Project AIR FORCE (PAF) to estimate the magnitude of potential savings that could accrue.[2]

1 A *BB contract* is a type of multiyear contract that is similar in many respects to the more formal multiyear procurement contract, but is subject to fewer regulatory requirements.

2 Although we offer general comments about potential areas of risk, we do not perform a formal risk assessment. Such an assessment would be needed to make a final decision about whether to pursue a BB.

Analysis Approach

This report summarizes PAF's assessment of cost savings available through a BB contract for production of three lots of F-35 aircraft—specifically, lots 12, 13, and 14, which are scheduled to be contracted for fiscal years (FYs) 2018, 2019, and 2020, respectively. To establish a baseline, we independently estimated the annual contracting costs of the air vehicle and engine under conservative, moderate, and aggressive assumptions. The scope of the estimate is limited to the recurring flyaway cost of the aircraft, which comprises airframe, mission systems, vehicle systems, engine, and engineering change orders. We next considered three categories of savings associated with BB contracts:

- **economic order quantities (EOQ),** in which Congress (or a partner nation) provides early funding so the contractor can leverage volume price breaks and other efficiencies to achieve cost savings
- **cost reduction initiatives (CRIs),** in which the contractor[3] invests in design or manufacturing improvements to reduce the per-unit cost of the product
- **administrative and other savings,** such as management challenges or fee reductions, as well as reduced burden on the supplier to bid, propose, and negotiate multiple contracts.

Drawing upon historical data, contractor and JPO estimates, and independent calculations (detailed in the body of this report), we identified a range of plausible cost savings in each category that would be available to an F-35 BB contract.

Using the above inputs, we constructed a model to estimate overall savings for the F-35 BB contract. The model begins with the annual contracting cost baseline and applies savings from each of the various

[3] We focus on contractor-funded CRIs that are expected to pay back within the period of the BB contract, but not within an annual contract. Government-funded CRIs may expect the investment to pay back over the life of the program, and are therefore not limited to BB contracts. However, because some multiyear contracts include government-funded CRIs in their savings estimates, we calculate net cost savings from the F-35 BB contract both with and without government-funded CRIs. Where we did not count government-funded CRIs as BB savings, we applied them to the annual contracting baseline.

categories to arrive at a BB contract cost. The model allows the following parameters to vary within ranges determined from the preceding steps as part of a Monte Carlo simulation:

- annual contracting baseline
- government and contractor funding for CRIs
- EOQ funding
- degree to which savings for suppliers who did not provide savings estimates are extrapolated.

The Monte Carlo simulation was run 5,000 times to arrive at the BB savings estimate and the range of likely savings.

Estimated BB Savings

Figure S.1 shows the results of these model runs as a frequency distribution of estimated savings, compared with RAND's estimated annual contracting baseline (results compared with the JPO's baseline estimate are also shown at the bottom). Because the F-35 air vehicle and engine are contracted separately, we show the results for each. The median savings are about $1.8 billion for the air vehicle and $280 million for the engine.[4] The median combined savings is about $2.1 billion, or 4.9 percent of the cost of procuring these lots through annual contracting, according to RAND's baseline estimate. These savings are roughly comparable to those estimated for historical multiyear contracts for other fighter aircraft.[5]

---

[4] The median is the middle value in an ordered list, such that there is an equal number of higher and lower values.

[5] Our savings estimate of 4.9 percent for the F-35 BB is lower than the 6.5 percent average multiyear savings estimates for F-16, F/A-18 E/F, and F-22 multiyear contracts. The gap closes when the savings arising from government-funded CRIs are added to our estimate. The F-16 and F-22 multiyear contracts did not include government-funded CRIs, but the F/A-18 E/F multiyear contract did.

Figure S.1
Distribution of BB Savings Estimates for F-35 Air Vehicle and Engine

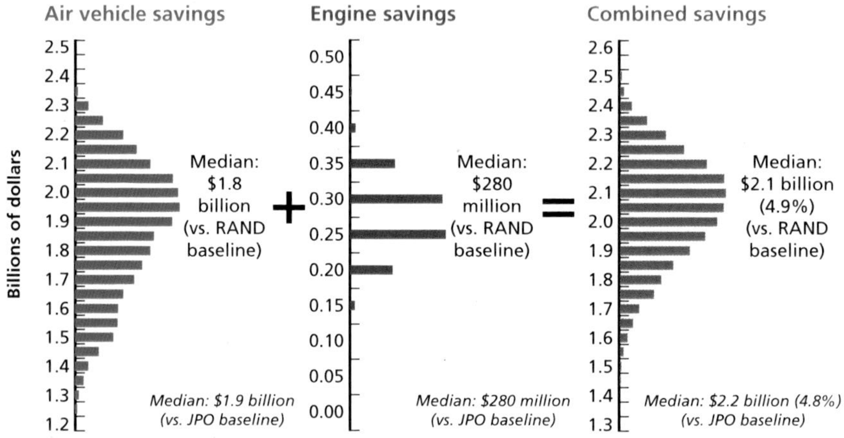

Air vehicle savings: Median: $1.8 billion (vs. RAND baseline); Median: $1.9 billion (vs. JPO baseline)

Engine savings: Median: $280 million (vs. RAND baseline); Median: $280 million (vs. JPO baseline)

Combined savings: Median: $2.1 billion (4.9%) (vs. RAND baseline); Median: $2.2 billion (4.8%) (vs. JPO baseline)

Billions of dollars

RAND *RR2063-S.1*

Hybrid BB: An Analytic Excursion

Owing to concerns about the availability of EOQ funding in FY 2017 at the time of this writing, we also assessed the potential savings of an alternative "hybrid" BB approach, in which a subset of countries would enter a BB contract for lots 12–14, with the remaining countries possibly entering the contract for only 13 and 14. We estimate the hybrid BB savings to be approximately 3.7 to 4.1 percent of the cost of contracting annually for the aircraft. These hybrid BB savings represent approximately 80 to 90 percent of the savings available in the original BB construct. The reduction in savings arises from a subset of countries deferring their commitment to the BB by one year—which, in turn, causes reduced economies of scale.

As an alternative to the hybrid BB approach, we also analyzed a simpler BB in which the United States and some partner countries are unable to make EOQ funding available in FY 2017. This approach achieves about 80 percent of the savings available in the original BB, but requires BB authorization for the United States and some partner countries a year earlier than in the hybrid BB approach.

Conclusions

Our analysis estimates potential savings from a BB contract for F-35 lots 12–14 to be approximately $2.1 billion for the air vehicle and engine combined. This is equivalent to 4.9 percent of the cost of annual contracting for these lots. This estimate is in the range of savings estimated for historical fighter programs that employed multiyear contracts.

PAF also considered alternative BB constructs, recognizing that EOQ funding in FY 2017 might not be available from all participating countries, including the United States. The estimated savings for these constructs is reduced by approximately 10 to 20 percent, compared with the full BB approach already described.

While we reiterate that this analysis focuses on *potential* cost savings from an F-35 BB contract and does *not* include a formal risk assessment, it provides bounding estimates that will inform relevant business decisions going forward. Potential areas of risk may include the availability of early EOQ funding, configuration changes, and aircraft quantity reductions. The JPO should consider these and other risks as part of its decisionmaking process and actively manage them if a BB contract is pursued.

Acknowledgments

The research and analysis presented in this report would not have been possible without the support of many individuals from the F-35 Lightning II JPO, as well as from Lockheed Martin, Pratt & Whitney, and their major suppliers.

We are grateful to Lt Gen Christopher Bogdan, Program Executive Officer for the F-35 JPO, for sponsoring this project, for providing excellent support for the project by opening many doors for us, and for engaging with and challenging us during several stimulating discussions related to this work. We very much enjoyed working closely with our JPO points of contact, Lt Col Michael Scales, Capt Kristin Ormaza, Maj Hakan San, and Lt Col Michael Dunlavy, each of whom did many things to ensure that this project was a success. Also, we are indebted to Anne Ryan, Michael Yeager, and Michael Clark at the F-35 JPO, who provided key cost data for our analysis and with whom we spent many hours discussing modeling approaches.

Many F-35 contractors were generous with their time in supporting this project. Foremost among them were the prime contractors, Lockheed Martin and Pratt & Whitney, who provided much support for this effort by supplying data, facilitating interactions with their major suppliers, and making themselves available to discuss details of the provided data and models. At Lockheed Martin, we wish to thank Jason Manning, Jim Glaub, and Brent Johnstone in particular; and at Pratt & Whitney, we wish to thank Tim Scanlon.

At RAND, we were fortunate to receive thoughtful feedback from many colleagues for this effort. Among them, we especially wish to thank Frank Camm, Natalie Crawford, Caolionn O'Connell, and Obaid Younossi, who greatly helped strengthen this work. Jerry Sollinger revised and edited an earlier draft of this document. Christina Hansen provided valuable administrative support throughout this project, and particularly with this document.

Abbreviations

AP	advanced procurement
AU	Australia
BB	block buy
CA	Canada
CAPE	Cost Assessment and Program Evaluation
CFR	Code of Federal Regulations
CIC	cost improvement curve
CRI	cost reduction initiative
CTOL	conventional take-off and landing
CV	carrier variant
DK	Denmark
DoD	U.S. Department of Defense
ECO	engineering change order
EOQ	economic order quantities
EVM	earned value management
FAR	Federal Acquisition Regulations
FFP	firm fixed-price

FMS	Foreign Military Sales
FPIF	fixed-price incentive (firm target)
FY	fiscal year
GAO	U.S. Government Accountability Office
IT	Italy
JPO	Joint Program Office
LCS	Littoral Combat Ship
LMA	Lockheed Martin Aeronautics
LRIP	low-rate initial production
NDAA	National Defense Authorization Act
MYP	multiyear procurement
NGAS	Northrop Grumman Aerospace Systems
NL	Netherlands
NO	Norway
P&W	Pratt & Whitney
PAF	Project AIR FORCE
ROM	rough order of magnitude
RR	Rolls-Royce
SAR	Selected Acquisition Report
STOVL	short take-off and vertical landing
TR	Turkey
TY	then-year
UK	United Kingdom
U.S.C.	U.S. Code

CHAPTER ONE

Introduction

The F-35 Lightning II is the most expensive program in U.S. Department of Defense (DoD) history. It is intended to replace and/or complement several fighter and attack aircraft for the U.S. military services, as well as those from a number of partner allied nations. Although the per-unit cost of the aircraft has been decreasing from lot to lot, the United States and partner nations are keenly interested in ways to reduce program cost further. One proposal is to procure three upcoming production lots under a single block buy (BB) contract, as opposed to multiple annual contracts. A BB contract is a type of multiyear contract that is similar in many respects to the more formal multiyear procurement (MYP) contract, but is subject to fewer regulatory requirements.[1] Like MYP contracts, BB contracts allow prime contractors and their suppliers to leverage quantity and scheduling certainty and economies of scale in order to implement cost-saving measures that would not be available when operating under serial annual contracts. However, multiyear contracts (be they BB or MYP) do not come without risks, which may include the availability of funding, configuration changes, and aircraft quantity reductions. To inform decisions about whether to pursue a BB contract for the upcoming production lots, the F-35 Joint Program Office (JPO) asked RAND Project AIR FORCE (PAF) to estimate the magnitude of potential savings that could accrue.

[1] We use the term *multiyear contract* as a generic term covering both formal MYP and BB contracts. Appendix A discusses the history of multiyear contracting and details the differences between BB and MYP contracts.

Program Background

The F-35 is a fifth-generation, single-seat, multirole fighter aircraft that employs stealth technology, fusing of sensor information, and network operations. It is being produced in three variants, designed for the U.S. Air Force, Marine Corps, and Navy and their partner-nation counterparts:

- F-35A is the conventional take-off and landing (CTOL) variant intended to replace Air Force F-16 fighters and A-10 attack aircraft.
- F-35B is the short take-off and vertical landing (STOVL) variant intended to replace Marine Corps AV-8B Harriers and F/A-18C/Ds.
- F-35C is the carrier variant (CV) intended to complement Navy F/A-18E/F fighters.

Participating countries are divided into tiers depending on their degree of financial commitment to program. Tier 1 includes the United States and the United Kingdom (UK). Tier 2 includes Canada, Italy, and the Netherlands. Tier 3 includes Australia, Denmark, Norway, and Turkey. In addition, Israel, Japan, and South Korea will acquire the F-35 through Foreign Military Sales (FMS).

The F-35 program includes a planned total of 2,457 aircraft for the U.S. Air Force, Marine Corps, and Navy. This includes 14 research and development aircraft and 2,443 production aircraft: 1,763 F-35As for the Air Force, 340 F-35Bs for the Marine Corps, and 340 F-35Cs for the Navy.[2] In addition, more than 700 aircraft—the vast majority being F-35As—are expected to be purchased by allied countries.

Given the size of the contract and the complexity of the aircraft, a large number of contractors are involved in the program, and they are spread across the United States and international partners' industrial

[2] The planned procurement quantities are always subject to change, and in fact were adjusted on multiple occasions while this work was being conducted. The numbers cited here are the ones used in this analysis and were current as of June 2016. The U.S. quantities can be found in Defense Acquisition Management Information Retrieval (DAMIR), *Selected Acquisition Report: F-35 Joint Strike Fighter Aircraft (F-35)*, Washington, D.C.: U.S. Department of Defense, December 2015.

bases. Lockheed Martin Aeronautics (LMA) is the prime contractor for the air vehicle, which comprises the airframe, mission systems, and vehicle systems. Other major air vehicle contractors are Northrop Grumman and BAE Systems. Pratt and Whitney (P&W) is the prime contractor for the F135 engine, which powers the F-35. Rolls-Royce (RR) builds the vertical lift system for the F-35B as a subcontractor to P&W.

While the F-35 program has had challenges, including cost overruns, developmental delays, and design issues, program costs have declined over recent years due to improving manufacturing costs and greater economies of scale from increased quantities.[3] Nevertheless, affordability remains an important concern, and the analysis presented here is motivated by this consideration.

Research Objective and Approach

This report assesses the potential savings that might arise from using a BB approach to three production lots of F-35 aircraft: lots 12, 13, and 14, which are scheduled to be contracted for fiscal years (FYs) 2018, 2019, and 2020, respectively.[4] These lots account for 471 aircraft, approximately half of which will be procured by the United States. Although we offer general comments about potential areas of risk, we do not perform a formal risk assessment. We recommend that such an assessment be included as part of the business case analysis that leads to the final decision about whether to pursue a BB.

The analysis included the following steps:

1. Establish a baseline estimate for the cost of procuring the aircraft for lots 12–14 through an annual contracting strategy. Consistent with how contracting is handled for the F-35 program, we developed separate annual contracting cost estimates

[3] Joint Strike Fighter Program, "F-35 Lightning II Program Fact Sheet: Selected Acquisition Report 2015 Cost Data," March 24, 2016.

[4] Procurement of a single aircraft lot involves activities that cover several years, from purchasing long-lead items all the way through delivery of the final aircraft. For convenience, we associate each lot with the planned year of the full-funding contract award.

for the air vehicle and the engine. In each case, we developed independent cost models that drew upon data from the prime contractors, their major suppliers, and the F-35 JPO.
2. Identify potential sources of BB savings and estimate the range of savings that could be expected for each. To develop these estimates, we drew upon the literature and regulations surrounding MYP and BB contracting, case studies of weapon system programs that previously employed MYP and BB contracts, and data from rough-order-of-magnitude (ROM) estimates developed by the prime contractors and their major suppliers.
3. Model potential savings against the annual contracting baseline to produce an estimated range of net savings.

In addition to the three-year BB analysis, we were asked to assess the savings associated with two somewhat more complex BB contracting strategies, in which a subset of international partner countries enters a BB for lots 12–14 and the remaining countries, including the United States, have the option to join the BB contract for lots 13 and 14. For these analyses, we used the results from step 1 above, and performed steps 2 and 3 using the ground rules associated with the new BB strategies and new data provided for this purpose by the prime contractors and their major suppliers.

All costs and savings reported in this document are presented in then-year (TY) dollars unless stated otherwise.

Report Organization

The remainder of this report is organized as follows: Chapter Two provides the annual contracting cost estimates, which form the basis of comparison for the BB contracts. Chapter Three describes our approach for estimating BB savings from individual sources. Chapter Four describes the methodology we used to integrate the savings estimating approaches and provides overall savings estimates. It also compares those savings to previous BB and MYP contracts. Finally, it outlines potential areas of risk that could challenge those savings and should be considered as part of a full business case analysis. Chapter Five describes the alternative BB contracting approaches and the savings associated with them. Chapter Six presents a summary of the analysis.

As additional context, Appendix A discusses the evolution of multiyear contracting and details the similarities and differences between BB and MYP contracts. Appendix B (published as a stand-alone document) examines case studies for nine historical weapon system programs that employed BB and MYP contracts.[5]

[5] Appendix B (RR-2063/1-AF) is a separate document available to the general public. An additional appendix of supporting information is not available to the general public.

CHAPTER TWO

Estimating the Annual Contracting Cost Baseline

In this chapter, we summarize our approach and present our estimate of the annual contracting cost for F-35 lots 12–14. This annual contracting cost is the baseline against which BB savings are evaluated. We present separate cost estimates for the air vehicle and engine, which is consistent with how contracting is performed in the program. In this unrestricted portion of the report, we do not describe details of the analysis that require presentation of contractor proprietary information.

Ground Rules and Assumptions

The production profile assumed in our analysis, current as of June 2016, is depicted in Figure 2.1. The F-35 JPO provided this production profile, and it is the same one underlying the 2015 F-35 Selected Acquisition Report (SAR).[1] As the figure shows, the total number of aircraft expected to be procured during the BB period is 471.

The scope of our estimate is limited to recurring flyaway cost of the aircraft, which comprises airframe, mission systems, vehicle systems, engine, and engineering change orders. These recurring flyaway cost elements are depicted in Figure 2.2, along with other procurement cost elements that are outside the scope of our analysis. We focus on recurring flyaway costs for two reasons:

[1] A *SAR* is a comprehensive summary of a Major Defense Acquisition Program that is required for periodic submission to Congress by the Secretary of Defense.

Figure 2.1
F-35 Program Aircraft Quantities Through the Proposed BB Period

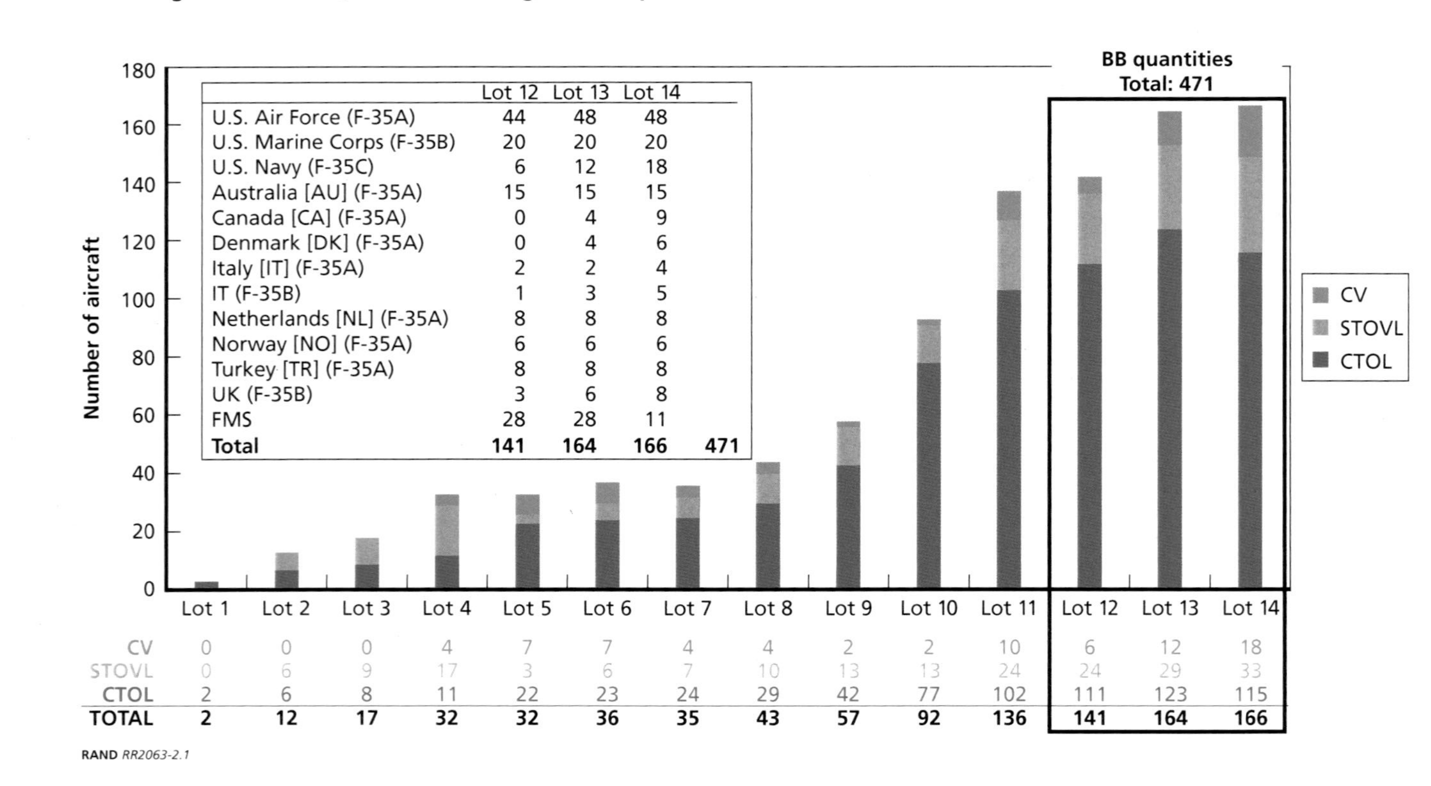

Figure 2.2
F-35 Procurement Cost Element Structure

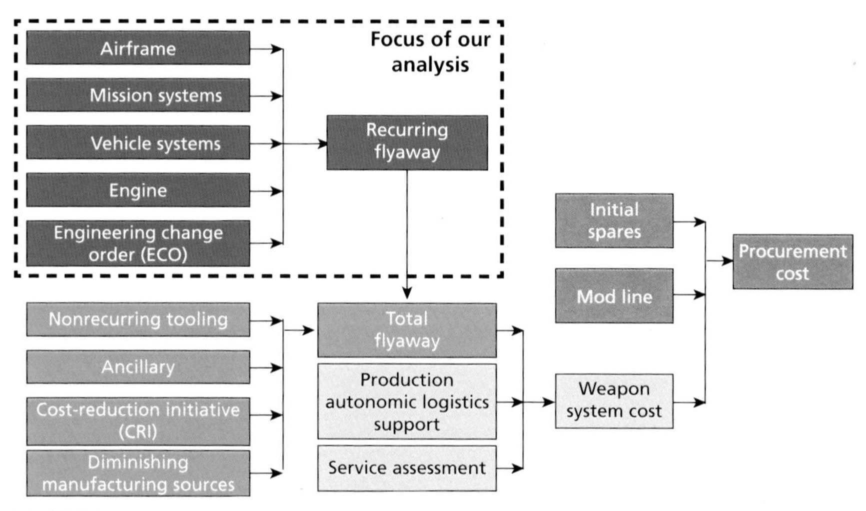

RAND *RR2063-2.2*

- Historically, approximately 70 percent of the procurement cost of the F-35 aircraft has been accounted for by the recurring flyaway cost.
- The annual contracting and BB savings data provided by F-35 contractors were limited to recurring flyaway costs.

Figure 2.3 provides additional detail on the composition of the F-35 recurring flyaway cost in our model. The airframe portion of the recurring flyaway cost estimate consists of in-house touch labor and support labor.[2] This includes all labor performed by the major air vehicle contractors indicated—LMA, Northrop Grumman Aerospace Systems (NGAS), and BAE UK—and not subcontracted out to other firms. In addition to in-house labor, each major contractor incurs general procurement costs, including the costs of raw material, purchased parts, standard hardware, and outside production. A large portion of the costs incurred by these major air vehicle contractors is major sub-

[2] *Touch labor* is work done by operators that directly adds value to the product. *Support labor* represents functions that enable production (e.g., program management).

Figure 2.3
Recurring Flyaway Cost Element Structure

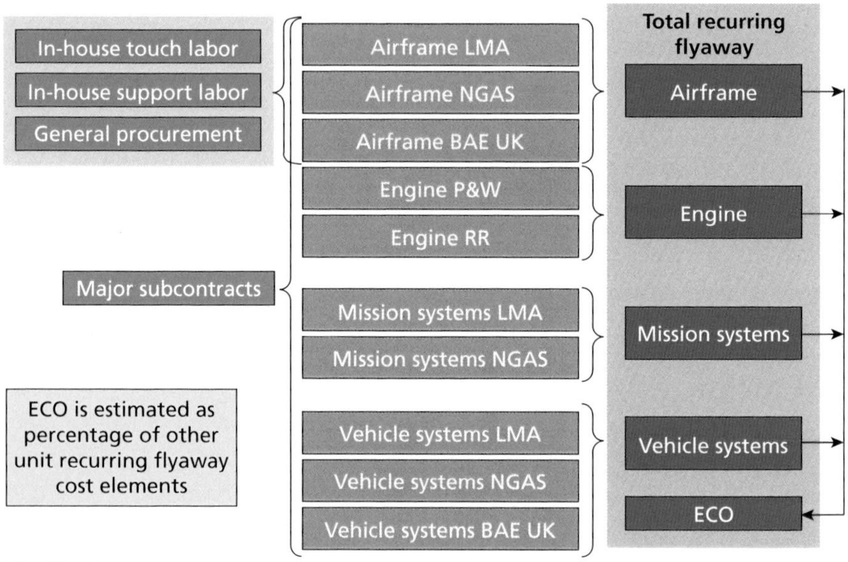

In-house touch labor
In-house support labor
General procurement
Major subcontracts
ECO is estimated as percentage of other unit recurring flyaway cost elements
Airframe LMA
Airframe NGAS
Airframe BAE UK
Engine P&W
Engine RR
Mission systems LMA
Mission systems NGAS
Vehicle systems LMA
Vehicle systems NGAS
Vehicle systems BAE UK
Total recurring flyaway
Airframe
Engine
Mission systems
Vehicle systems
ECO

RAND *RR2063-2.3*

contracts. LMA and NGAS have major subcontracts for mission systems, and LMA, NGAS, and BAE UK have major subcontracts for vehicle systems. In addition to subcontracts for mission systems and vehicle systems, all of the major air vehicle contractors have major subcontracts to offload a portion of their airframe work. These major subcontracts might include coproduction with foreign firms or other divisions within the firm (e.g., LMA produces part of a wing in its Fort Worth operations and another part in its Marietta operations).

When available, we used inflation indices provided by the contractors to inflate or deflate their cost data. When indices from the contractors were not available, we used indices provided by the F-35 JPO that were tailored to commodity type and region/country.[3]

[3] The JPO inflation indices were produced by an economic consulting firm, Global Insight. These indices are adjusted or weighted by various factors, including commodity types and regions/countries. Due to the global production and various commodity types used to produce the F-35, these indices were judged to be the most appropriate where contractor indices were not available.

For the three major airframe contractors, we obtained labor rates and material burdening rates from the F-35 JPO and judged them to be reasonable. These rates account for costs related to fringe, manufacturing and engineering overhead, program office support, general and administrative expenses, facilities capital cost of money, and fees, and are based on forward rate-pricing agreements and forward rate-pricing proposals between the government and contractors.[4]

Finally, note that cost improvement arising from CRIs is accounted for in our estimates through the historical data on which these estimates are based.[5] In other words, a sustained level of cost improvement is implicitly assumed as we project costs forward from historical data that inherently reflect some level of cost improvement from CRIs. The exception to this is a government CRI investment of $300 million over lots 12–14, which is assumed to generate cost improvement beyond historical levels.

The following sections detail our data sources, estimation methodologies, and resulting estimates for the air vehicle and engine.

Air Vehicle Baseline Estimate

Data Sources

For the air vehicle, we collected and used data from several sources, including the major contractors (LMA, NGAS, and BAE UK) and the F-35 JPO. In the majority of instances, data from lots 4–7 were used because prior lot data were not available for the air vehicle. In most instances, projections for labor hours (both touch and support labor) were based on actual hours provided by the JPO. However, there were several lower-level cost elements for which contractor-

[4] As will be described in the following methodology discussion, anywhere labor hours are used in regressions and labor rates are applied, overhead costs are explicitly calculated and included in the labor rates. However, for items that are subcontracted or coproduced (e.g., mission systems, vehicle systems, coproduced items) and for data that are reported as "material dollars," labor and associated overhead costs are already included, so we do not apply any additional overhead costs.

[5] A detailed definition and discussion of CRIs is presented in Chapter Three.

provided hours were used. In cases where labor projections were based on realization factors (to be defined later in this section), contractors provided industrial engineering standards. For mission systems, negotiated values for lot 8 that were provided by the JPO were used as a starting point for regressions, while actual costs of vehicle systems provided by the contractors were used in the regressions. Offloaded quantity assumptions were based on the contractors' most recent plans, when available. For instances in which the contractor data for offloads were not provided, the latest JPO in-house percentage assumptions were used. Finally, as mentioned earlier, burdened labor rate and material estimates for raw material, purchased parts, standard hardware, and outside production are based on data provided by the JPO.

Labor

The cost-estimating community has used several methods to determine in-house touch labor hours. One of the most common methods is using a regressed trend of historical hours data to generate a cost improvement curve (CIC).[6] However, this method can lead to inaccurate estimates in the presence of changes to configuration, tooling, manufacturing methods, CRI investment, or customer/market pressures. An alternative method that is more robust to such program changes relies on the use of realization factors (RFs). An RF is defined as the ratio of the actual hours to complete a task to the task's industrial engineering standard,[7] and can thus be viewed as a metric that normalizes data to mitigate the aforementioned complications. We used an approach that employs RFs in conjunction with CICs to estimate in-house touch labor, when reliable data were available to

[6] A *CIC*, also known as a *learning curve*, refers to the constant rate of reduction in cost for each doubling of quantity or production rate, assuming no major changes in product design, production processes, workforce composition, and interval between units.

[7] An *industrial engineering standard* is the time taken by a qualified operator to perform a given task on a repetitive basis, including fatigue and delay allowances.

do so.[8] When we could not use the RF approach because industrial standards data were not available, we often developed CICs using actual hours.[9]

To account for commonality across the three F-35 variants, we analyzed touch labor for each variant using its own CIC in which the RFs or hours for that variant were regressed against cumulative *program* quantities (i.e., all variant quantities); the exceptions to this are sections or components that are variant-unique. The rationale for using cumulative program quantities (as opposed to variant quantities) is that manufacturing tasks (e.g., drilling holes; sealing; driving fasteners; installing tubes, harnesses, and other equipment items; testing) should be performed with the same efficiency across the variants as the program matures.

Support labor was estimated as a percentage of in-house touch labor, and this percentage was informed by F-35 and legacy aircraft program experience.

Material

For material cost elements (e.g., raw material, purchased parts, standard hardware, and outside production), we used contractor data to develop estimates. However, in cases where contractor data were inadequate to do so, we used JPO-provided estimates instead.

One of the high-value areas of the material estimate is *coproduction*, which in this context refers to instances where two or more firms are producing the exact same section of the aircraft. We independently calculated coproduction estimates, which include the labor and material of the coproduced part of the aircraft. Some examples of coproduction include LMA coproducing sections of the wing with Alenia and Israel Aerospace Industries; NGAS coproducing the inlet duct, center

---

[8] Specifically, the CIC approach we used was ordinary least squares regressions of the natural logarithm RF against the natural logarithm of the cumulative lot midpoint quantity for various sections of the aircraft (e.g., forward fuselage, wing, center fuselage, mate). In instances where RF data were not available or reliable, our CIC regressions consisted of the natural logarithm of actual in-house touch labor hours against the natural logarithm of the cumulative lot midpoint quantity.

[9] Other approaches (e.g., a percentage factor applied to another cost element) were used to estimate costs for a handful of other cost elements.

fuselage subassembly, and center fuselage assembly with Turkish Aerospace Industries; and BAE UK coproducing portions of the vertical and horizontal tail assemblies with Marand and Magellan. All coproduction estimates assume a "fully competitive price": the amount the government will pay for coproduction will be no greater than the cost to produce the product by the major air vehicle contractor (i.e., LMA, NGAS, BAE UK), less any supplier burdens. It is assumed that any additional costs incurred by the coproducing firm above the major contractors' cost would be subsidized by the local government and these subsidies are not included in our estimate. Therefore, the labor hours for coproducing firms are the same as the hours for the prime contractor. In addition to the labor portion of the coproduced aircraft sections, the material costs need to be accounted for. To estimate the material portion of the coproduction costs, we determined a ratio of material to labor based on actuals from prior lots and applied this factor to the labor estimates.

Mission Systems and Vehicle Systems

Because mission system configurations have been changing with progressive block upgrades, we did not use CICs based on actuals. Instead, we formed estimates using the latest negotiated value and a slope based on legacy program experience from that point forward.[10] Mission systems were assumed to be common across all three variants.

Vehicle systems were estimated by regressing actuals from completed lots.[11] Vehicle systems were categorized as either common among all three variants or variant-unique. Common vehicle systems include those that may be similar among variants but are not necessarily identical. Common vehicle systems actuals for CTOL aircraft were summed and regressed against total program quantities. Vehicle system costs for STOVL and CV aircraft were estimated by applying a

---

[10] Ronald Smouse and Paul Tetrault, *Joint Strike Fighter Avionics Cost Improvement Study (Learn, Rate, Step-Down, and Other Considerations)*, F-35 Joint Program Office, November 2002.

[11] Specifically, we used ordinary least squares regressions of the natural logarithm of the average vehicle system unit cost (in constant year dollars) to the natural logarithm of the cumulative lot midpoint quantity.

factor, based on actuals, to the CTOL estimate. For variant-unique systems, actuals were summed up for each variant and regressed against variant quantities.

Developing an Estimate Range

After developing a point cost estimate for the air vehicle, uncertainty in the following key modeling parameters led us to consider alternative values for them, which generated a range in our estimate:

- in-house touch labor CIC slopes
- mission systems CIC slopes
- coproduction CIC slopes
- support labor percentages.

Results

Tables 2.1 and 2.2 summarize the results of all the calculations using the approach described for the air vehicle. Table 2.1 summarizes the results on a per-aircraft basis, whereas Table 2.2 summarizes the results on a per-lot basis using the production profile in Figure 2.1.

Table 2.1
Air Vehicle Estimated Average Unit Recurring Flyaway Costs for Annual Contracting (millions of TY dollars)

	CTOL			STOVL			CV		
Lot	Low	Mid	High	Low	Mid	High	Low	Mid	High
12	68.7	73.6	78.1	80.5	82.3	86.5	89.5	93.4	97.5
13	66.4	71.4	76.4	78.2	79.8	84.3	84.2	88.8	93.1
14	65.4	70.5	76.4	76.7	78.1	82.8	81.5	86.5	91.1

Table 2.2
Air Vehicle Estimated Total Recurring Flyaway Costs for Annual Contracting (billions of TY dollars)

Lot	Low	Mid	High
12	10.1	10.7	11.3
13	11.5	12.2	13.0
14	11.5	12.2	13.2

Figure 2.4 illustrates our estimates of the annual contracting baseline cost for the air vehicle, compared with the JPO's estimate underlying the 2015 F-35 SAR.[12] The RAND range on the annual contracting baseline cost is represented by the blue bar associated with each lot, with a point estimates (i.e., mid estimate) indicated by the horizontal white line across the bar. The green dot is the JPO estimate from the FY 2015 SAR.

The dominant driver explaining why the RAND estimate is lower than the JPO estimate is that our in-house touch labor estimates for LMA, NGAS, and BAE Systems are lower. The RAND estimate relies on prior lot actuals and analogies to historical aircraft programs to estimate touch labor costs. Moreover, because touch labor estimates form the basis of support labor and coproduction cost estimates, the effect of lower touch labor estimates is compounded through these other cost elements. Overall, the RAND point esti-

Figure 2.4
Annual Contracting Cost Baseline for the Air Vehicle (billions of TY dollars)

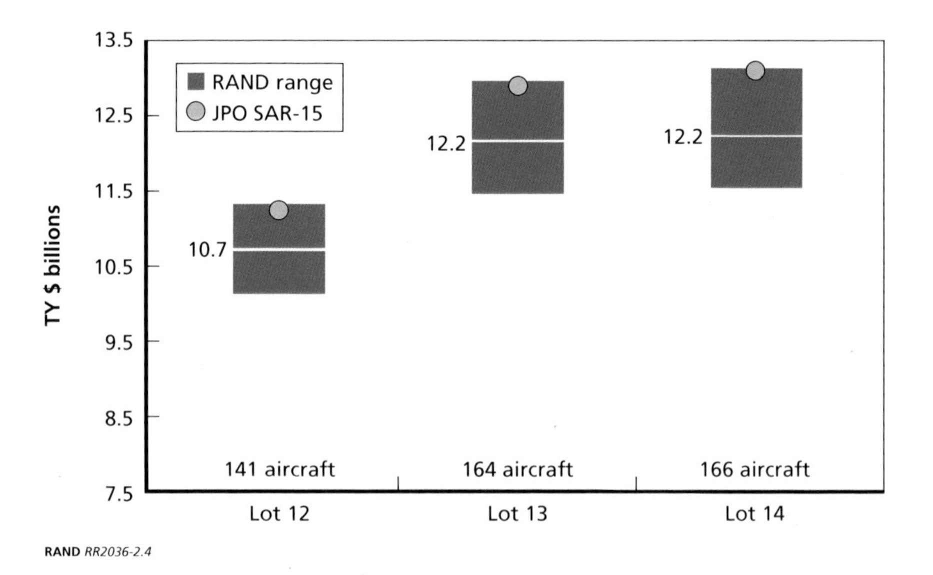

[12] While the 2015 SAR for the F-35 only reports production quantities for the United States, a full production profile including non-U.S. countries is needed to estimate aircraft costs. The 2015 SAR production profile assumes 471 F-35 aircraft over the BB period.

mate is approximately 5.6 percent lower than the JPO's. As a result, BB savings estimates will be commensurately lower when they are computed relative to the RAND annual contracting baseline versus the JPO's (as discussed in Chapters Four and Five).

Engine Baseline Estimate

Data Sources

We collected data on the actual cost of engines and engine components produced by P&W and RR. The data provided by P&W and RR cover all engines produced until partly into lot 8. In addition, we collected Cost and Software Data Reporting (CSDR) data from each contractor and the JPO.[13] P&W provided a file with manufacturing costs by Component Integrated Product Team (CIPT) and for the whole engine (but excluding the lift fan system produced by RR for the STOVL engine). The engine modules covered by the CIPTs include the fan, high-pressure compressor, diffuser/combustor, high-pressure turbine, low-pressure turbine, turbine exhaust case and augmentor (afterburner), nozzle, mechanical systems, controls, externals, assembly and test, 3-bearing swivel module (STOVL engine only), and standard hardware. The lift fan system for the STOVL engine is made by RR and makes up approximately half of the overall costs of a STOVL engine. RR provided cost data for engines in lot 2 through the first four engines in lot 8. These are actual costs from RR's cost performance reports.

Estimating P&W and RR Portions of Engine Cost

We use regression analysis on the engine cost data provided to us by P&W and RR. As already discussed, the P&W data cover the entire cost of the

[13] The CSDR data cover the charges to the government of each contract and reflect various contracting and negotiating outcomes that result in differing numbers of engine elements being covered by each contract and dollar values that represent the results of negotiations, which in some cases do not reflect the actual costs of the engines in specific annual buys. The purpose of the present analyses is to establish an estimated baseline for the actual cost of single-year procurement of engines for lots 12–14, so we decided to rely on the engine-by-engine cost data provided by the contractors instead of the CSDR data.

CTOL/CV engine variant and the cost of the STOVL engine, excluding the lift fan system. For the STOVL engine, we add the projected cost of RR's lift fan system to P&W's portion of the STOVL engine cost to arrive at the full cost of the engine. All engine costs projected using the regression analysis were then adjusted upward to account for general and administrative expenses, facilities capital cost of money expenses, and fees to arrive at average unit recurring flyaway costs for the engine.

Estimating P&W's Portion of the Cost of Engines

For P&W's portion of engine costs, we utilize cost and other information for 211 engines produced as part of lots 1–8. We regress the engine costs reported in P&W's data on the unit number (to capture learning effects), the quantity of CTOL/CV and STOVL engines in each engine's lot (to capture any economies of scale in the production of engines), and whether the engine was CTOL/CV or STOVL.[14] As part of the model formulation, we tested alternative regressions specifications, which enabled the rate of learning captured by the unit number to vary for earlier and later lots (i.e., allowing for a "break point" in the learning slope). We also considered regressions run on alternative subsets of the available data (i.e., lots 1–8 compared with lots 6–8). During discussions with P&W, it was indicated that significant design changes went into effect between lots 5 and 6. As a result, we allowed for breaks in the learning curve effect between those lots. We also used the model to explore data from only lots 6–8.

Estimating the Cost of the RR Lift Fan System for STOVL Engines

RR's cost data covers 52 lift fan systems used in the STOVL engine. We estimated the cost of the lift fan system using a regression specification that controls for the unit number to capture learning effects.[15] As

---

[14] Specifically, we ran ordinary least squares regressions of the natural logarithm of P&W's reported engine cost versus the natural logarithms of the engine unit number and the quantity of CTOL/CV and STOVL engines in each engine's lot, plus a dummy variable for the STOVL engines.

[15] The regression relates the natural logarithm of the cost of RR's lift fan systems to the natural logarithm of the lift fan system unit number.

with the P&W costs, we developed alternative estimates by performing a regression on different subsets of the data (lots 2–8, 4–8, and 5–8).

Developing an Estimate Range

We develop low, mid, and high estimates for the engine based on the relative ranking of projected costs for the regression specifications examined.

- The low estimate is derived from projections based on lots 6–8 for P&W and lots 2–8 for RR.
- The mid estimate is derived from projections based on lots 1–8, with a learning break between lots 5–6 for P&W and lots 4–8 for RR.
- The high estimate is derived from projections based on lots 1–8 for P&W and lots 5–8 for RR.

Results

Tables 2.3 and 2.4 summarize the results of all the calculations using the approach described. Table 2.3 summarizes the results on a per-engine basis; Table 2.4 summarizes the results on a per-lot basis using the production profile in Figure 2.1.

Table 2.3
Engine Estimated Average Unit Recurring Flyaway Costs for Annual Contracting (millions of TY dollars)

	CTOL/CV			STOVL		
Lot	Low	Mid	High	Low	Mid	High
12	12.0	12.9	13.3	29.1	31.5	31.9
13	11.7	12.7	13.1	28.9	31.7	32.1
14	11.7	12.7	13.2	29.1	32.0	32.6

Table 2.4
Engine Estimated Total Recurring Flyaway Costs for Annual Contracting (billions of TY dollars)

Lot	Low	Mid	High
12	2.1	2.3	2.3
13	2.4	2.6	2.7
14	2.5	2.7	2.8

Figure 2.5
Annual Contracting Cost Baseline for the Engine (billions of TY dollars)

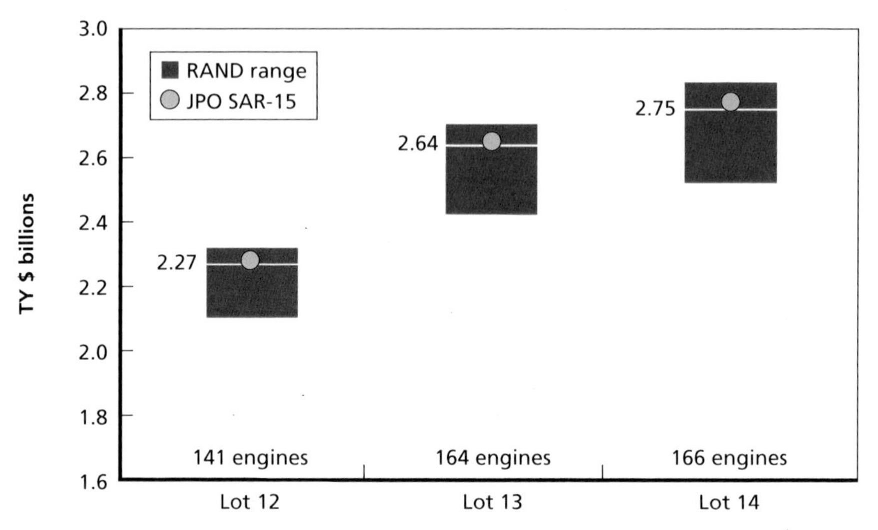

RAND *RR2036-2.5*

Figure 2.5 illustrates our estimate of the annual contracting baseline cost for the engine, compared with the JPO's estimate underlying the 2015 F-35 SAR. The RAND range on the annual contracting baseline cost is represented by the red bars associated with each lot, with point estimates (i.e., mid estimates) indicated by the horizontal white lines across the bars. The mid RAND and JPO estimates are quite similar; however, the rapid cost improvement from lot 6 to lot 8 drives a low estimate for the low end of the RAND estimate range.

Summary

We described our approach and estimates of the annual contracting costs for the F-35 air vehicle and engine. We also compared our estimates to those of the JPO and found them to be within our range. The annual contracting costs estimated in this chapter serve as the baseline against which BB savings are evaluated in the remainder of this report. As discussed in Chapter Four, we model BB cost savings against the full range of baseline estimates.

CHAPTER THREE

Assessing Potential BB Savings

In this chapter, we describe the approach and methodology for assessing potential BB savings for the F-35 program. We first present some background on the characteristics and incentives associated with annual and multiyear contracts, and identify some key assumptions for our analysis. Then we present our methodology for estimating each component of potential F-35 BB savings.

Military Procurement Contracting

DoD contracting for most weapon systems, including the F-35, occurs in an unusual environment, where the government is the sole buyer of the system and there is a single contractor that is the sole source of the system.[1] This monopsony/monopoly situation has many implications for program costs and contract price negotiations that have been discussed at length in the literature.[2] Here, we discuss only a small subset

[1] The F-35 consists of two systems, the air vehicle and engine, each of which is procured in this manner by the U.S. and partner country governments.

[2] See, for example, Jean Tirole, "Procurement and Renegotiation," *Journal of Political Economy,* Vol. 94, No. 2, 1986; William E. Kovacic, "Commitment in Regulation: Defense Contracting and Extensions to Price Caps," *Journal of Regulatory Economics,* Vol. 3, 1991; William P. Rogerson, "Economic Incentives and the Defense Procurement Process," *Journal of Economic Perspectives*, Vol. 8, No. 4, Fall 1994; and Chong Wang and Joseph San Miguel, "Unintended Consequences of Advocating Use of Fixed-Price Contracts in Defense Acquisition Practice," *Proceedings of the Eighth Annual Acquisition Research Symposium, Wednesday Sessions*, Vol. 1, Naval Postgraduate School, 2011.

of the issues associated with military procurement that are of particular interest. These include the issues of contract type, price negotiation approach, and incentives for generating cost savings.

Contract Type

Recent low-rate initial production (LRIP) of lots of the F-35 has been performed under a series of annual fixed-price incentive (firm target) (FPIF) contracts, one of many contract types available for military procurement, as described in the Federal Acquisition Regulations (FAR).[3] Since there have been no formal negotiations between the government and major contractors around a potential BB contract as of fall 2016, there is some uncertainty about the type of contract that will be used for the BB. Some of the contractors and suppliers, in preparing their BB savings estimates, assumed the BB would be conducted under a firm fixed-price (FFP) contract, in which case they would not be subject to earned value management (EVM) requirements.[4] Relaxing EVM requirements can save money in the near term, but reduces the flow of timely information on program cost and status to the program office. The F-35 JPO communicated to RAND that it expects to use an FPIF contract for the BB, which would include the same EVM requirements that have been included in the LRIP contracts. We proceed with this assumption in our analysis.

As specified in the FAR, "A fixed-price incentive (firm target) contract specifies a target cost, a target profit, a price ceiling (but not a profit ceiling or floor), and a profit adjustment formula."[5] The nature of an FPIF contract, then, is that the final cost to the government is not necessarily the same as the target cost ("price") written into the contract. If the contractor's actual costs overrun or underrun the target cost in the contract, the profit adjustment formula is used to determine the actual amount paid to the contractor. The objective of our work

---

[3] 48 Code of Federal Regulations (CFR) §16.403-1, 1994.

[4] EVM can be applied to an FFP contract, but is "discouraged" and a waiver must be obtained to do so. See DoD, Defense Federal Acquisition Regulation Supplement (DFARS), Procedures, Guidance and Information (PGI), Section 234.201, "Policy," December 7, 2011.

[5] 48 CFR §16, 1994.

is to estimate the cost savings to the government that are potentially available from pursuing a BB: We are not deriving the cost target that would be used in the resulting contract. In fact, because our analysis includes some level of contractor investment that must be recouped, achieving our estimated savings requires that the target cost in the contract be higher than the expected government cost. The size of that differential depends on the profit adjustment formula, but our savings estimate does not.

Savings Incentives

Because of the sole-buyer and sole-seller nature of the F-35 program, the BB or target cost cannot be determined through competitive market forces. Rather, the price is based on cost estimates of the type described in Chapter Two, with a profit margin for the contractor added on. Both the cost estimates and profit margin are negotiated between the government and prime contractor(s). This situation creates a number of challenges for the government as it attempts to simultaneously negotiate a reasonable cost target and incentivize the contractor to generate savings.

First, the FPIF contract is designed to reward a contractor that reduces costs below the target during contract execution. This feature can incentivize contractors to seek cost savings, but it also incentivizes them to negotiate the highest possible cost target so that it is easier to underrun.[6] Future cost information is inherently asymmetric: The contractor is in a much better position to estimate its future production costs and savings opportunities than the government is. It can thus be difficult for the government, which is eventually privy to actual costs incurred, to determine whether cost underruns arose from good-faith CRIs or overstated initial cost estimates.

However, the government is not completely blind during negotiations. It has access to actual cost data from prior production lots, if there were any. Also, the Truth in Negotiating Act (TINA) requires that contractors and large subcontractors for sole-source noncompetitive proposals submit certified cost or pricing data as part of their pro-

---

[6] Wang and San Miguel, 2011.

posals.[7] However, the contractor is not obligated to provide the lowest possible cost at which it could complete the procurement. TINA encourages contractors to make their actual costs approximate the negotiated costs as closely as possible. In fact, if a subcontractor generates significant savings during contract execution, it could face punishment in the auditing stage if the government concludes that the cost underrun was actually a result of fraudulent initial cost estimates.[8] The FPIF contract, where savings are shared between the government and contractor through the profit adjustment formula, does mitigate these concerns to some extent.

Second, while the contractor may benefit in a single FPIF contract from reducing costs thanks to the profit-adjustment formula, future contracts are again negotiated based on estimated costs. If a cost-saving measure generates persistent savings such that future lots cost less than they would have otherwise, the cost target of future contracts and therefore the contractor profit on those contracts will be similarly reduced. The contractor may then defer cost-saving measures until later lots nearer the end of the program. Further, if such a cost-saving measure requires investment on the part of the contractor, and that investment is not paid back before the contract period, the contractor is likely to lose money on that measure—again, because future contracts are based on the lower cost that the cost-saving measure permitted. These effects create a disincentive for the contractor to pursue the kinds of cost-saving measures that the government would most desire: those that bend the cost curve through many program lots rather than just during a single contract. One method used to mitigate this disincentive is for the government to invest in CRIs. But the government is still dependent on the contractor to identify CRIs—which a contractor may be reluctant to do, given that CRIs reduce future costs and associated profits.

---

7 10 U.S.C. (U.S. Code) 2306a, 2011; 41 U.S.C. Chapter 35, 2011. Subcontractors whose part of the contract is more than $700,000 are subject to TINA requirements. A more extensive discussion of negotiation on military defense multiyear contracts can be found in Appendix A.

8 Kovacic, 1991.

Depending on the program, there may also be meaningful incentives for contractors to reduce costs. If contractors envision future commercial or foreign sales of the weapon system, it may be important to them to reduce costs to make the system more competitive in the marketplace. Lower costs can also make the government less likely to reduce procurement quantities or terminate the program early. Finally, major defense contractors must consider future programs on which they will likely bid. There is value in maintaining good relations with DoD and Congress, and in avoiding the bad press associated with contracts that appear to be bloated.

For the purposes of our analysis, we make the following assumptions about contractor incentives. Given the history and visibility of cost overruns in the F-35 program, we believe that the contractors are more than sufficiently incentivized to reduce costs in the long run. Thus, we assume that contractors will not be averse to CRIs or other measures that persist beyond the BB period, lowering the cost of future lots. However, we also assume that contractors will not invest their own money to implement a CRI unless they can obtain a sufficiently positive return on that investment within the BB contract period. These assumptions will be discussed further later in this chapter when we estimate CRI savings. Finally, since we are developing and using our own cost estimates for the annual and BB contracts, we make no assumption about the baseline costs that will eventually be negotiated between the government and the contractors. We are estimating cost savings that we perceive to be available in a BB contract, not the cost savings that the government will actually manage to negotiate, which may be lesser or greater.

Sources of BB Savings

Multiyear and BB contracts potentially enable savings beyond those that are available under annual contracting. Natural learning and rate increases generate per-unit savings regardless of the contract type or length, as can be seen in the analysis in Chapter Two. To determine the savings associated with BB contracting, we must identify savings that require some form of multiyear contracting to achieve. Historically, prenegotiation estimates of MY and BB contract savings have ranged widely, from around 4 to 14 percent in programs we examined (see Appendix A for details). Consistent with prior analyses, we were unable to identify significant trends in the data on previous programs that would meaningfully bound a savings estimate for the F-35 BB. Therefore, we focused on the potential sources of BB savings and estimated the benefit of each one for the F-35 program. In the end, we consider savings in three broad categories, distinguished by the type of funding required to enable savings (if any) and the nature of savings achieved. These categories are (1) economic order quantities (EOQ), (2) CRIs, and (3) administrative and other savings. We discuss each of these categories in the remainder of this section. Analysis of the potential savings is only one piece of information that goes into the BB decisionmaking process. See Appendix A for a discussion of the legal requirements and congressional expectations for MYP and BB contracting.

EOQ

EOQ usually refers to how companies optimize the quantities of material they order to minimize holding and ordering costs, while enabling optimally efficient production planning and operations. This takes into account volume price breaks and the costs of shutting down and restarting production lines. An extensive literature on EOQ exists,[9] although only a small fraction of studies have evaluated EOQ in the context of military procurement.

In the context of most DoD procurement programs, funding is allotted on a single-year basis. Contractors can order parts more or less frequently than once a year to fulfill production requirements, but if

[9] See, for example, Ford W. Harris, "How Many Parts to Make at Once," *Factory: The Magazine of Management*, Vol. 10, No. 2, 1913; R. H. Wilson, "A Scientific Routine for Stock Control," *Harvard Business Review*, Vol. 13, No. 1, 1934; J. A. Buzacott, "Economic Order Quantities with Inflation," *Operational Research Quarterly (1970–1977)*, Vol. 26, No. 3, 1975; Franklin Lowenthal, "Cost of Prediction Error in the Economic Order Quantity Formula," *Managerial and Decision Economics*, Vol. 3, No. 2, 1982; and S. K. Goyal, "Economic Order Quantity Under Conditions of Permissible Delay in Payments," *Journal of the Operational Research Society*, Vol. 36, No. 4, 1985.

more than a year's worth of inputs are being ordered or produced, they face some risk of cancellation losses or rework if the procurement requirement is terminated or changed under single-year contracting. When procurement is pursued under a multiyear or BB arrangement, the risk of termination or obsolescence is reduced, and special provisions enable the government to provide funding to support cost savings through EOQ.[10]

EOQ is a component of advanced procurement (AP) funding, which allows procurement "a fiscal year in advance of that in which the related end item is to be acquired."[11] To be clear, this is money that is already within the procurement budget; it is simply provided earlier than it would be under annual contracting. EOQ funding is generally limited to recurring costs and therefore cannot be used to fund CRIs. As noted by the Congressional Research Service,[12] authority to use EOQ differs under MYP and BB contracting. Specifically, authority to provide EOQ funding does not come automatically as part of BB contracting authority; Congress must specifically authorize EOQ funding.

EOQ savings are generally considered to be a major source of MYP or BB savings. Note, however, that EOQ savings are not persistent: They save money within the contract period, but they do not reduce the per-unit cost of the product in the lots that come after the BB.

Determining how much EOQ funding a government should authorize under a BB contract is complex. It depends on the expected

[10] The extent to which defense contractors formally employ EOQ methodologies to determining order quantities is unclear. Nevertheless, funding to expand procurement opportunities beyond a single year has been termed "EOQ funding" in DoD contexts. Our analysis focuses on EOQ savings enabled by government funding. In theory, contractor funding of EOQ can also generate savings for the government. However, to receive a portion of these savings, the government would have to negotiate this agreement and possibly provide a termination liability in the event that the contract is cancelled. This issue of contractor funding of EOQ is revisited in the next two chapters.

[11] As noted in Appendix A, the other component of AP funding is called "long-lead funding," which allows for the procurement of parts and material one year in advance if they are required to maintain a planned production schedule. General Services Administration, Federal Acquisition Regulation, Part 217, acquisition.gov website, January 19, 2017.

[12] Ronald O'Rourke and Moshe Schwartz, *Multiyear Procurement (MYP) and BB Contracting in Defense Acquisition: Background and Issues for Congress*, Washington, D.C.: Congressional Research Service, R-41909, March 4, 2015.

savings that each dollar of advanced procurement will return, on the timing of the EOQ money and those savings, and on competing demands on a constrained government budget. Obtaining the first year of EOQ funding (which will generally be provided in the year before the first lot year of the contract) can be a particular challenge. EOQ funding in later years will be offset by the savings that have been generated and shifts of EOQ funding out of those contract years, lessening the impact on the federal budget. We make no attempt in this work to evaluate how much EOQ funding the governments should provide. The F-35 JPO has stated that it expects to provide 4 percent of the contract value in the form of EOQ funding, 2 percent each in advance of the first two contract years. Therefore, we assume 4 percent EOQ funding is available in this analysis.

CRIs

CRIs are activities whereby a company, with an up-front investment, changes a design or process to reduce the per-unit cost of the product. This up-front investment, which is eventually recouped through the savings brought about by the CRI, is not part of the cost of production and therefore not associated with future production lots; it is extra funding that must be provided over and above ongoing production expenses.

CRI savings are, or at least can be, persistent. If the actual per-unit cost of a component is reduced, those savings should be enjoyed even in lots beyond the BB contract, perhaps until the end of the program, but at least as long as that component or process is used in production.

The per-unit savings associated with a CRI are available regardless of whether the government or the contractor funds the up-front investment. The savings to the government associated with a CRI will generally be quite different, however. If the government funds a CRI, the full savings associated with that CRI, from implementation through the end of the program, should accrue to the government. Therefore, the government's willingness to fund CRIs should be largely independent of the contract length. As a result, while savings from government-funded CRIs are real and valuable, we do not consider them to be BB-specific savings. (However, as discussed in Appendix B, programs have often historically included government-funded CRIs in their BB and MYP savings estimates. We take this into account in Chapter Four, when we compare total savings from an F-35 BB contract to estimates from previous programs that had multiyear contracts.)

If a contractor is investing its own money to fund a CRI, however, the contractor will recoup a fraction of the savings generated during the contract period as additional profit, with the fraction determined by the profit adjustment formula in the FPIF. At the end of the contract, any future savings generated by the CRI should accrue to the government, as the per-unit cost reductions now become part of the baseline cost estimate. Thus, in deciding whether to invest its own money in CRIs, the contractor cares a great deal about the contract length. Multiyear contracts produce greater regulatory lag, a longer period over which the contractor can take CRI savings as additional profits before they are negotiated away in the next contract.[13] Note that regulatory lag is not inherently bad for the government: incentivizing the contractor to gain profits through CRI investment benefits the government in the current FPIF to the extent of its share of the profit adjustment formula, and via lower future costs, assuming the program continues beyond the current contract. Also, the government can leverage the existence of regulatory lag to negotiate a lower target price, essentially challenging the contractor to discover and implement CRIs in exchange for obtaining the multiyear contract and retaining some of the additional savings as profit.

Given the fact that government-funded CRIs will return more savings for the government than contractor-funded CRIs, one might expect that the government would be willing to fund all CRIs that it determines have a positive return on investment. However, as with the advanced procurement money that supports EOQ savings, the amount of government money available for CRI investment may be constrained. The F-35 JPO informed us that it anticipated having an upper limit of $300 million in CRI investment money available for the

[13] For an excellent discussion of regulatory lag and the associated incentives, see Rogerson, 1994.

BB, $100 million in advance of each of the three BB contract years. We assume this level of funding availability in our analysis.

Administrative and Other Savings

This category includes BB savings that do not require additional funding (as with CRIs) or early funding (as with EOQ), but are nevertheless available in the BB contract. These include, for example, administrative savings that arise from proposing and negotiating three lots at once, rather than having three separate contract proposal and negotiation activities. There are a number of items included in this category, which will be listed later in this chapter. These savings generally are not persistent: The program will benefit from them only during the BB contract period. In some cases, regulatory lag could enable these savings; activities that would not be cost-effective in an annual contracting environment may prove worth doing under a multiyear contract. Thus, the contractor is incentivized to discover and implement these measures during contract execution, keeping a share of the savings as additional profits.

Management Challenge and Fee Reduction

This important subcategory captures savings that a contractor offers to the government that are not tied to any specific reductions in production costs. Contractors may offer these savings if they perceive great benefit to obtaining the contract. MYP and BB contracts can offer benefits to contractors of multiyear business stability and the potential to obtain additional profits through activities related to regulatory lag as described above. Thus, a contractor may be willing to offer a management challenge, wherein they promise savings before having identified their source. Or the contractor may agree to reduce the standard fee, accepting a lower profit in exchange for the benefits associated with a multiyear contract.

In the ROM savings estimates we received, a few, but not many, contractors and suppliers offered management challenge savings and fee reduction savings. In our analysis, we evaluated the potential for savings in this category based on an analysis of the benefits of regulatory lag. We assume that the contractor will be incentivized to attempt to find savings during contract execution that will allow it to meet the challenge or make up the lost fee amount. Conversely, contractors will only offer these savings to the government if they are confident that they can make up the difference during contract execution. In general, these activities will take the form of contractor-funded CRIs, where some investment is required to bring in savings. We therefore evaluate the potential for savings in this category as part of our CRI analysis, and our methodology and results for this savings category are described in the CRI section later in this chapter.

The three categories of savings and their key characteristics are summarized in Table 3.1.

The remainder of this chapter is organized as follows. In the next section, we briefly discuss the data sources that were of greatest use in our analysis. That section is followed by a detailed description of the methodology employed for each of the three categories of savings listed in Table 3.1.

Data Sources

We examined a number of sources to help assess savings in these three categories. We explored the relevant literature and analyses of previous MYP and BB contracts for weapon system programs, with the objective of identifying predictive relationships for different sources of savings. However, we found that these data sources did not yield useful relationships that could be used for our purposes because, as noted in Appendix B, significant differences among key program characteristics

Table 3.1
Block-Buy Savings Categories

Savings Category	Persistent Savings	Funding Required	Funding Type
EOQ	No	Yes	Advanced procurement
CRI	Yes	Yes	Up-front investment (government or contractor)
Administrative and other	No	No	N/A, except management challenge and reduced fee, where cost recovery occurs through contractor-funded CRIs

require that each program be evaluated in depth with respect to its unique characteristics. This finding was consistent with prior analyses of multiyear contract savings.[14]

Thus, for our BB savings analyses, we primarily drew upon data that F-35 prime contractors and their suppliers provided. One important source of data was the ROM estimates of savings that contractors formulated in summer 2015 in response to a request from the F-35 JPO. However, given the limited amount of time and resources available to the contractors for this effort, the ROM estimates do not have the same level of fidelity as contract proposals or negotiated contracts. RAND, therefore, requested that contractors improve the fidelity of their estimates—for example, by separating BB savings that require EOQ funding from those that do not. These data were key to estimating savings in the categories of EOQ and of administrative and other savings. To estimate CRI savings, we obtained and analyzed both historical F-35 CRI data and contractor estimates of CRI savings that would be available during the BB period.

We scaled our final results to show the savings estimate against other baseline cost estimates, such as the JPO estimate. In Chapter Four, we use the full range of RAND annual cost estimates as part of a Monte Carlo simulation to determine a range of likely savings. Whenever results of this analysis are reported, it is important to highlight the annual cost baseline against which the BB savings were evaluated.

[14] See for example V. Sagar Bakhshi and Arthur J. Mendler, *Multiyear Cost Modeling*, Fort Lee, Va.: Army Procurement Research Office, 1985, and Obaid Younossi, Mark V. Arena, Kevin Brancato, John C. Graser, Benjamin W. Goldsmith, Mark A. Lorell, Fred Timson, and Jerry M. Sollinger, *F-22A Multiyear Procurement Program: An Assessment of Cost Savings*, Santa Monica, Calif.: RAND Corporation, MG-664-OSD, 2007.

Estimating Savings from EOQ Funding

This section describes our approach to estimating BB savings from EOQ funding. The defense literature generally categorizes savings generated from EOQ funding into three areas:[15]

- **large buys of parts and material.** Buys of parts and material are increased beyond what would be necessary to support single-year procurement, thus exploiting economies of scale and quantity discounts.
- **production build-out and acceleration.** Production rates for major components are increased beyond what would be necessary to support single-year procurement, thus exploiting economies of scale.
- **support labor.** Support labor encompasses activities such as production planning, engineering, tooling support, supplier management, financial analysis and reporting, cost estimating and pricing, and contract administration. A reduction in production costs or an acceleration of supplier production and deliveries can lead to a decrease in support labor costs.

When government EOQ funding is available, it is generally left up to the prime contractor(s) to determine how that funding is allocated to its own activities and to its suppliers. To achieve the greatest level of savings, the government would prefer that EOQ funds be distributed efficiently (i.e., in such a way that they maximize savings for a given level of government funding) rather than some other scheme, such as evenly allocating funding among major suppliers. To distribute EOQ funds efficiently, the prime contractors must first understand how returns on EOQ differ across EOQ funding opportunities. The EOQ funds should then be distributed to the opportunities with greatest savings per dollar of EOQ provided. In addition to allocating EOQ funding efficiently, the government would prefer that the prime contractors avoid using EOQ funding for items at risk of future

[15] See Younossi et al., 2007.

design changes, which could render prepurchased items unusable or require costly rework. This may limit the set of items that prime contractors consider eligible to receive EOQ funding or require higher savings thresholds for certain components at greater risk of future design changes.

Seeking a Predictive Relationship Between EOQ Funding and Savings

We sought a way to predict potential savings from EOQ funding by examining how the above considerations have borne out in historical aircraft MYP and BB programs. The primary sources of information used in this analysis were the budget justifications submitted to Congress for these programs.[16] These data are summarized in Table B.3 in Appendix B.[17]

Some programs publish multiple budget justifications. In these instances, we use the last available budget justification in our analysis. The information we collected from budget justifications was supplemented with information from program office visits and other sources when available. Data on program savings generated from the F-22 MYP contract were never published. For this particular program, we used savings and other program information contained in previous RAND research.[18]

Figure 3.1 plots the estimated percentage of total estimated savings relative to total contract value against the percentage of each pro-

---

[16] The budget justification relied on here should be viewed with some caution. See Kathleen P. Utgoff and Dick Thaler, *The Economics of Multiyear Contracting*, Washington, D.C.: Center for Naval Analyses, 1982. Also, note that the military services have incentives to overstate savings from MYP contracts. Furthermore, there is limited effort to validate the savings estimates published in the budget justifications; see U.S. Government Accountability Office (GAO), *DoD's Practices and Processes for Multiyear Procurement Should Be Improved*, Washington, D.C., GAO-08-298, 2008. Bakhshi and Mandler, 1985, discuss challenges generally with quantifying savings from MYP arrangements.

[17] Information from many of the budget exhibits were compiled and utilized in Younossi et al., 2007.

[18] Younossi et al., 2007.

Figure 3.1
EOQ Funding Request and Savings Percentages from Budget Justifications

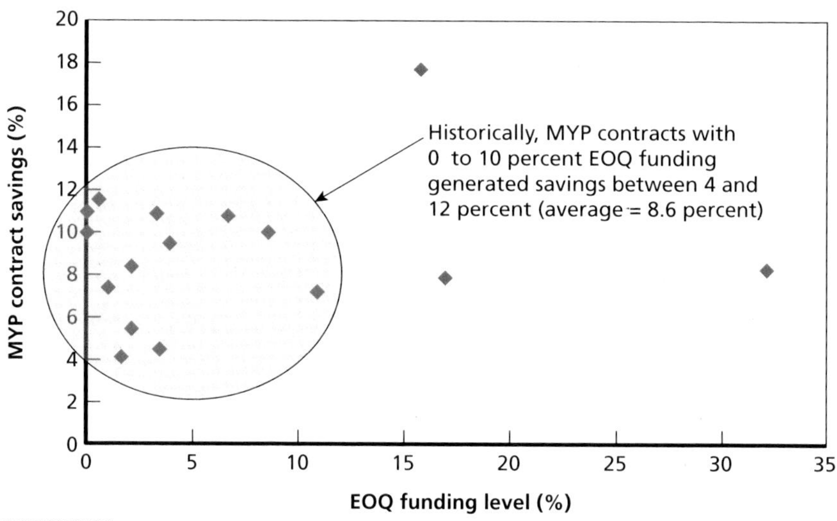

RAND *RR2063-3.1*

gram's EOQ funding request relative to total contract value.[19] We see that while programs that had EOQ funding realized some savings, there is no discernable correlation between the amount of EOQ funding and the resulting savings. This is true even when we control for other MYP contract characteristics, such as program size, era, or military service, using multiple regression techniques.[20]

---

[19] Total estimated savings is an imperfect measure, but we were unable to distinguish how much of the total savings was tied to EOQ activities. Ideally, we would be plotting *EOQ-enabled savings*, rather than overall savings, versus EOQ funding request, but these data were not available.

[20] In fact, we found almost no systematic relationship between MYP contract savings and any program characteristics, as discussed in Appendix B. Our finding is consistent with Younossi et al., 2007, which found that "for multiyear programs after 1995, we could find no significant statistical correlation between the contract savings estimates and such factors as contract size, total number of aircraft procured, number of aircraft procured annually, length of the MYP contract, or funding provided for economic order quantity (EOQ) or cost reduction initiatives" (p. xix). This is not surprising, given that our analysis is using the data from Younossi et al., supplemented with information on MYP contracts that occurred subsequently.

Table 3.2
Relationship Between Changes in EOQ Funding and Changes in Savings for Programs with a First and Second MYP Contract

	MYP I		MYP II		Change	
Program	EOQ (%)	Savings (%)	EOQ (%)	Savings (%)	EOQ (%)	Savings (%)
E-2C	32.2	8.3	10.9	7.2	–21.3	–1.1
V-22	1.6	4.1	0.5	11.6	–1.1	7.5
F/A-18E/F	1.0	7.4	0.0	10.9	–1.0	3.6
F-16	7.0	7.7	2.1	8.4	–4.9	0.7
CH-47	0.0	10.0	0.0	10.0	0.0	0.0
C-130J	3.3	10.9	3.9	9.5	0.6	–1.4
C-17	2.1	5.5	6.6	10.8	4.5	5.3

To explore further whether EOQ funding levels are correlated with savings, we looked at changes in EOQ funding and savings for aircraft programs with more than one MYP contract. Table 3.2 shows changes in EOQ funding and estimated savings as percentages of contract value for programs that pursued a first and second MYP contract. In general, we observe no statistical relationship between changes in EOQ funding percentages and changes in savings percentages for programs with multiple MYP contracts.

Failing to discern a useful predictive relationship for EOQ savings from previous MYP contract data, we next turned to the ROM estimates provided by major air vehicle and engine suppliers of the EOQ funding required and the related savings available in the F-35 BB under consideration. We examined these data in depth, attempting to find correlations between supplier characteristics, such as company size, contract size, part complexity, component type (e.g., electronic vs. mechanical), and savings. We were unable to identify any significant trends that would allow us to predict EOQ savings.

Modeling Savings from EOQ Funding

In the absence of an empirical basis for predicting EOQ savings based on historical experience, we used the following method to model potential savings in an F-35 BB contract.

If participating governments were able to make unlimited EOQ funding available, they would give every supplier all of the requested EOQ funding, while demanding the associated savings.[21] However, in reality, governments will be limited in the amount of EOQ funding they can make available. As discussed earlier in this chapter, the F-35 JPO anticipated that the U.S. and partner governments would provide up to 4 percent of the contract value in EOQ funding. To determine the potential savings available within a given level of EOQ funding, we used the EOQ savings estimates provided by the suppliers. First, we removed suppliers who requested FY 2016 EOQ funding, as there was no chance of such funding being available, after confirming with the suppliers that they indeed required the money in that year in order to achieve the savings they reported. Next, we ordered the submissions by the expected return on EOQ funding (dollars of EOQ savings offered divided by dollars of EOQ funding requested).[22] In our model, we then provided EOQ funding to suppliers in this order, working our way down the list, providing funding and claiming savings, until we reached the EOQ funding ceiling. This approach generates the curve shown in Figure 3.2, where cumulative EOQ funding (as a percentage of combined air vehicle and engine contract costs) is plotted on the horizontal axis and cumulative savings obtained on the vertical axis.[23] Note that the slope of the curve decreases from left to right, reflecting the fact that each additional dollar of EOQ funding garners decreases from left to right, reflecting the fact that each additional dollar of EOQ funding garners less savings than the previous dollar. In the end, 4 percent EOQ funding captures more than 80 percent of the available EOQ savings.

---

[21] This is true as long as the return offered by the supplier exceeds the governments' cost of capital. If this criterion is not met, the governments should not provide the supplier its EOQ funding request.

[22] Since not all suppliers were surveyed during the initial ROM data collection, we extrapolated the data we received to include all major suppliers. For air vehicle and engine, the data we received covered approximately 83 percent of the total cost to the governments of major supplier costs. We assumed in our analysis that data from the remaining suppliers would mirror the reported data.

[23] We do not show the individual suppliers in Figure 3.2 because the data are proprietary.

Figure 3.2
EOQ Savings vs. EOQ Funding Requested

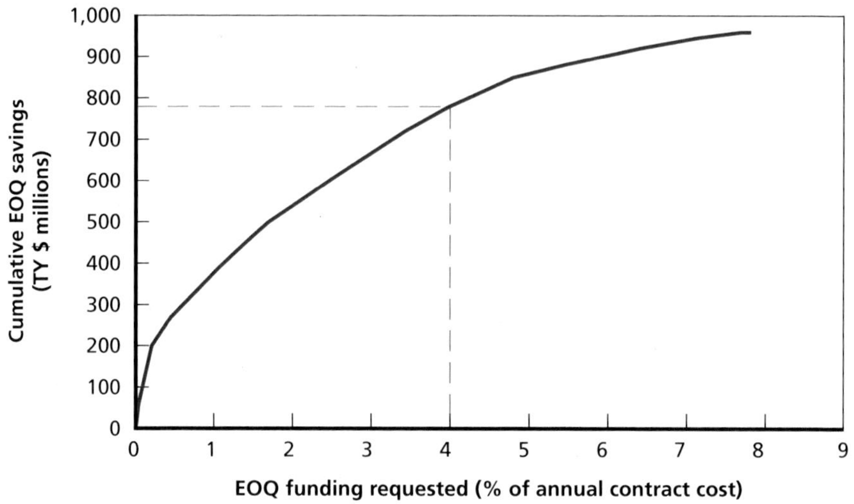

RAND *RR2036-3.2*

Estimating Savings from CRIs

This section describes our approach to estimating BB savings from CRIs. As discussed previously, CRIs can be funded by either the government or the contractor. The associated per-unit cost reduction will be the same in either case, but the amount of cost savings obtained by the government will differ. The F-35 JPO has said that $300 million in government funding will be made available in lots 12–14 for investment in air vehicle CRIs, $100 million per year. LMA took this investment into account in their ROM submission, performing a top-down estimate of the savings that would be achieved through this investment. There was no significant contractor CRI investment included in the ROM savings estimates we received.

Savings from Government-Funded CRIs

In its original ROM submission, LMA provided an estimate of the savings that it could achieve with the projected $300 million of government CRI investment. Based on some additional data provided by LMA describing how CRI savings are phased in over time, we constructed a model for the air vehicle that matches LMA's results; further detail is not described in this unrestricted portion of the report to protect the contractor's proprietary data underlying the model. This model has some key features that are worth noting. First, the rate of return on CRIs decreases as investment increases, which is not surprising; we would expect the most cost-effective CRIs to be the first ones implemented. Second, these savings decrease gradually, such that even the last of the $300 million of government investment yields a positive return on investment over the life of the program for any reasonable assumption about the government cost of capital. Finally, the implementation of CRIs is gradual, such that associated savings are small in the first year of implementation and grow in later years. This behavior, which means that it takes time for the CRI investments to pay back, limits the amount of savings accrued within the BB period, which has implications both for the BB savings we report (which ignores savings beyond the BB time horizon) and for the viability of contractor-funded CRIs, which will be discussed later in this chapter.

To assess whether the LMA model of CRI savings was reasonable, we obtained data on contractor-proposed CRIs in the F-35 program. These were primarily identified under the Blueprint for Affordability (BFA) and War on Cost (WOC) initiatives.[24] We evaluated the savings associated with these initiatives and the trends in these savings over time and found that the LMA model was very consistent with that data. Therefore, we accepted the LMA model and used it to evaluate the savings arising from government-funded CRIs.

Recall that we do not consider government-funded CRI savings to be BB-specific. In fact, the F-35 JPO plans to supply this CRI funding no matter what contracting approach is chosen for lots 12–14. Therefore, we apply this savings equally to our annual cost estimate and our BB cost estimate.

---

[24] BFA is an agreement among DoD and LMA, NGAS, and BAE Systems to invest $170 million from 2014 to 2016 in CRIs. The plan requires the contractors to make the initial CRI investments, which are recouped, with profit, after the cost reductions to the program have been validated. WOC is P&W's program of CRIs for the F-35 engine—akin to BFA for the air vehicle—which began in 2009.

Management Challenge and Contractor-Funded CRIs

The ROM savings estimates received from the contractors included essentially no contractor-funded CRIs. After government-funded CRIs are implemented, if additional CRIs are available that provide significant returns within the BB period, we would expect the contractor to implement these initiatives and claim the savings as additional profit. The government, knowing this, should be able to negotiate some amount of management challenge into the contract, enough for the government to receive some of the savings from these contractor-funded CRIs.[25] As long as the contractor believes it can meet the management challenge and still obtain a sufficient rate of return on these investments, it should be willing to agree to these terms in order to obtain the BB contract. For the purposes of this analysis, we assume that the annualized rate of return on investment that a contractor requires to implement a CRI is the same as the standard fee (profit margin) included in the contract. We further assume that the government will collect all savings from contractor-funded CRIs beyond that amount. Whether the government truly collects all these savings depends on the details of contract negotiations.

Note that the above logic could apply to contracts of any length, including annual contracts. Thus, BB-specific savings are available from contractor-funded CRIs only if there are CRIs that take longer than one year and shorter than three years to pay back. CRIs that pay back within one year would be available under annual contracting, and the contractor will not fund those that take longer than the BB contract period to pay back. Recall, also, that there is already $100 million per year of government-funded CRI taking place: We are looking for CRIs that meet these conditions and are still available after that government investment. We now turn to the question of how much savings from contractor-funded CRIs is potentially available to the government during the BB.

Ideally, CRIs with a positive net present value to the government and/or contractor should be implemented as early as possible in

[25] In our model, we treat this as a management challenge, but the same effect could be achieved in an FPIF contract by negotiating a more government-friendly share line in the profit adjustment formula.

an acquisition program. In reality, at least two constraints limit the amount of total CRI investment in any given year:

- At some point, limitations in the supply of skilled engineering personnel and/or excessive disruption to ongoing production would limit the number of CRIs that can be effectively executed at any given time.[26]
- The number of CRIs available that actually pay back within the appropriate time period may be limited.

To address the first question, we examined previous MYP and BB programs that included government-funded CRIs. When programs have included government-funded CRI investments, the annual investment has ranged from roughly 1 to 2.5 percent of the contract value. We do not have data on additional contractor-funded CRI investments made on those contracts, but we expect that there were some. We have used a conservative baseline assumption in our analysis that 2 percent of the yearly contract value could be put to work for government- and contractor-funded CRIs collectively in each year of the BB period. In our Monte Carlo analysis of savings (described in Chapter Four), we allow this investment level to vary more widely, from no contractor investment to a combined 4-percent investment. The $100 million of planned annual government investment is roughly 0.8 percent of the yearly contract value. We assume the remaining funds (approximately 1.2 percent of the yearly contract value) are available for the contractor to invest. This still leaves the question of whether there are CRIs available that justify this investment.

To evaluate this second question, we again used the historical F-35 CRI data referenced earlier. An individual CRI could be represented as a point on the pot in Figure 3.3, where we show the expected per-unit savings on the y-axis and the required investment to implement the CRI on the x-axis. We have divided this pot area into four sectors, related to the payback time associated with the CRIs. In this

[26] Another constraint on implementing government-funded CRIs is that the government does not always fund investments with positive net present value at its official interest rate because of budget and borrowing constraints.

Figure 3.3
Contractor-Proposed CRIs on the F-35 Program

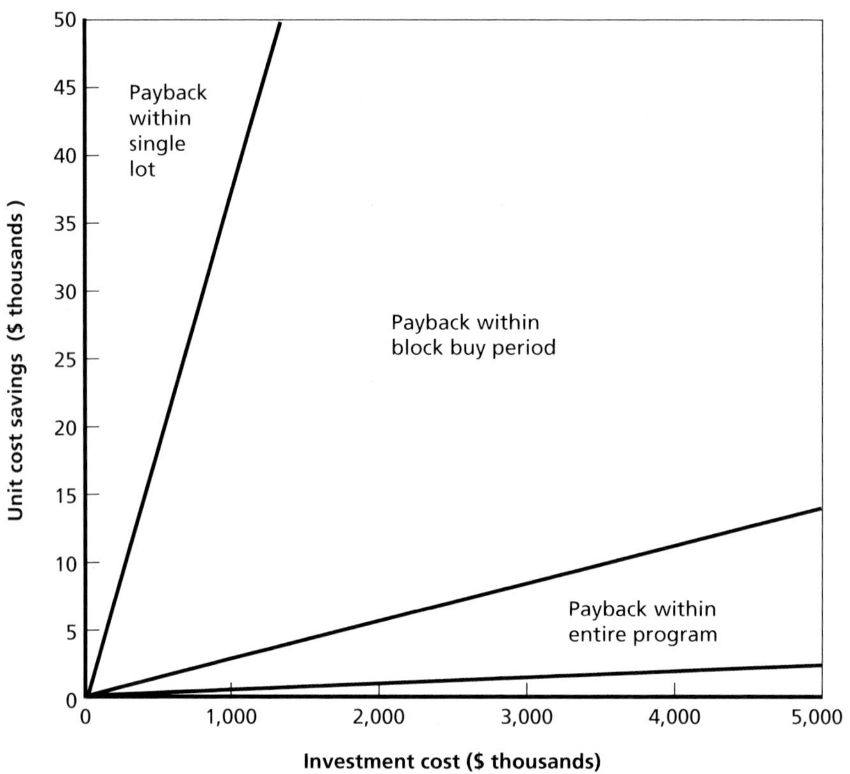

RAND *RR2063-3.3*

plot, the payback time for the single lot and for the BB is calculated for contractor investment, which means that the CRIs do not pay back until the contractor has recouped the initial investment plus the cost of capital (assumed to be the same as the profit margin included in the F-35 contract).[27] Moving clockwise from the top of the figure:

---

[27] The payback periods are also based on the F-35 production planning profile from the 2015 SAR, including international partners and FMS, and assume a gradual phase-in of CRI savings over time. The slopes of these lines would shift based on the quantities of aircraft produced in a single lot and during the BB period, as well as assumptions about the phase-in of savings.

- CRIs in the top sector provide a sufficiently high return that they pay back within a single lot. CRIs in this sector do not require a BB to be worthwhile investments for the contractor.
- CRIs in the large center sector, which we will call the "BB sector," pay back within one to three years (i.e., the BB period beyond the first year). The contractor should be willing to fund these CRIs if, and only if, an MYP or a BB contract is in place.
- The third section encompasses CRIs that need more time than the BB period allows for payback, but would pay back over the entire program. While the contractor would not be willing to fund these, the government might, as long as there is confidence that the program (or the aircraft components undergoing the CRI) will not change so much in the coming years that the benefits of the CRI will go away.
- The small sliver on the bottom covers CRIs that will not pay back over the current F-35 production profile. Note that it is possible for a CRI in this section to be worthwhile, perhaps because it improves reliability or safety or reduces sustainment costs. (The sectors in the figure are based on procurement cost analysis only).

Most air vehicle and engine CRIs for which we have data fall within the BB sector. When we examine CRIs over time, it appears that even very recently proposed CRIs fit primarily in this sector. This aggregation suggests that there is an ample supply of CRIs that would generate BB savings.

Modeling Savings from Contractor-Funded CRIs

Using the CRI savings model that we developed using LMA-provided data, we plotted the savings as a function of investment in a pot similar to Figure 3.3. The CRI returns generated by this model lie within the BB sector of Figure 3.3. The rate of return on CRIs decreases as investment increases, which is not surprising; we would expect the most cost-effective CRIs to be the first ones implemented. However, the savings decrease gradually, such that if we extrapolate the model beyond the $300 million government investment, we find that the savings remain within the BB sector until investment reaches well beyond the

2-percent limit we impose.[28] These results validate our approach and suggest that contractors ought to be willing to fund these CRIs because they would be able to recoup their investments plus their profit margin, while providing savings to the government.

To develop a model for engine CRIs, we combined two pieces of information. First, from an analysis of air vehicle and engine CRI data, we know that engine CRIs have historically offered quite similar returns on investment to air vehicle CRIs. From this, we concluded that the LMA model should be an appropriate approach for engine CRIs as well. Second, we performed a regression of historical engine CRIs, scaled the LMA model to match our engine analysis, and then applied this modified model to evaluate contractor-funded engine CRIs in the BB period.

A few companies did include management challenge and/or fee reduction in their submissions. Even though these are not exactly the same thing, we assume in both cases that the contractor will attempt to recover their full fee during the contract, and our model of contractor-funded CRIs is accounting for that. Therefore, when we impose our model of contractor-funded CRIs onto the contractors, we proportionally reduce any management challenge and fee reduction savings they offered, to avoid double-counting savings.

Estimating Administrative and Other Savings

This category covers BB savings that do not require extra funding (as CRIs do) or early funding (as EOQ savings do), but are available under any BB or MYP contract. The following types of savings appear in the F-35 contractor BB ROM submissions:

- **reduced bid, proposal, and negotiation of contracts.** For multiple annual contracts, supplier efforts in support of bid development, proposal development, and contract negotiations must be

---

[28] In fact, this is true all the way to the 4-percent investment limit we use in our Monte Carlo analysis discussed in Chapter Four.

repeated for each contract. Only a single iteration of these activities is required for a BB contract, resulting in reduced costs.

- **reduced EVM reporting.** Periodic EVM reporting requires that suppliers regularly provide detailed program management reports to the program office. The manpower cost of developing these reports can often be significant. As such, eliminating or reducing the periodicity of EVM reporting can reduce program costs, if it is determined that more-frequent reports are not necessary.[29]
- **reduced supplier and material management.** A BB contract would reduce the manpower requirements associated with supply chain and material storage management.
- **engineering change proposals funded separately.** Some F-35 suppliers suggested funding engineering change proposals by a contract separate from the F-35 BB contract.[30] This would reduce costs for the F-35 BB contract, but would increase costs for a different government contract.
- **foreign exchange rate management.** For suppliers whose supply chain relies on multiple currency types, fluctuating exchange rates present a variable cost risk that is subject to unpredictable volatilities. This risk can be reduced through up-front agreements with banking organizations that offer bulk "lock-in" of exchange rates to significantly reduce the impact of currency fluctuations on program costs. In some cases, this risk reduction can result in cost savings to a program.
- **long-term agreements with suppliers.** A BB contract would allow suppliers to guarantee business for their subtier suppliers for a longer period of time. This provides the opportunity for suppliers to negotiate lower costs with their supply chain.
- **build-rate adjustments.** Given the larger total number of aircraft guaranteed to be manufactured in a BB contract (as opposed to in multiple separate annual contracts), suppliers and their subtier

---

[29] Per DoD acquisition regulations, EVM reporting is required for all non-FFP contract types and can be relaxed at the discretion of an individual program office.

[30] An *engineering change proposal* is a contractor-recommended configuration change that comes with an associated cost.

vendors can optimize production rates to maximize efficiency and achieve cost savings.

- **alternative sourcing.** Guaranteed procurement of a larger number of aircraft can enable subcomponents or subassemblies to be purchased from alternative sources. Suppliers may solicit competitive bids for subcomponents, which might not be a cost-effective strategy for a single year's procurement. Alternative sourcing also includes outsourcing, or purchasing traditionally in-house developed subcomponents or subassemblies from external sources.
- **capital investments.** Like CRIs, capital investments can offer cost savings by improving program efficiency in a variety of ways. They are funded through a contractor's capital budget and depreciate over time.
- **management challenge and reduced fees.** *Management challenge* refers to a supplier setting a cost-savings goal, independent of other savings initiatives, in order to reduce program costs. In some cases, suppliers may be willing to accept a reduced fee or profit in a BB contract environment as compared with an annual contract environment. This is because the BB contract's long-term guarantee of business improves long-term profit certainty.

Treatment of Proposed Savings Sources

Table 3.3 summarizes the different sources of savings proposed by the F-35 contractors and how we treated them in our BB savings analysis. Savings ideas with a checkmark in the "BB Only" column were judged to be available only through a BB contract and were thus counted as BB savings in our analysis. Savings ideas with a checkmark in the "Annual and BB" column were judged to be legitimate savings to the program, but available in both annual and BB contracting environments and therefore not eligible for BB savings in our analysis. Finally, savings ideas with a checkmark in the "No Savings" column were judged not to be available to the F-35 program in any contracting environment.

Table 3.3
Treatment of Proposed BB Savings Sources

Category	Initiative	BB Only	Annual and BB	No Savings
EOQ	Buyout of parts and material	✓		
	Production build-out and acceleration	✓		
	Support labor savings	✓		
CRI	Contractor funded	✓		
	Government funded		✓	
	Implemented prior to lot 12		✓	
Administrative and other	Reduced-bid, proposal, and negotiation of contracts	✓		
	Reduced EVM reporting			✓
	Reduced supplier and material management	✓		
	Engineering change proposals funded separately			✓
	Foreign exchange rate management	✓		
	Long-term agreements with suppliers	✓		
	Build-rate adjustments	✓		
	Alternative sourcing	✓		
	Capital investments	✓		
	Management challenge and reduced fee	✓		

As the table shows, we consider EOQ and contractor-funded CRIs as true BB savings. However, as discussed earlier, government-funded CRIs do not need a BB contract to implement, and so these savings would be available in an annual contracting environment. Therefore, in this analysis, we applied the savings from government-funded CRIs to the annual contracting baseline. In addition, CRIs implemented before the BB period, proposed by some contractors as BB savings, were not counted toward savings because they will be implemented regardless of whether a BB contract happens. Most of the savings ideas in the "Other" category were judged to be legitimate BB savings. The exceptions were:

- "reduced EVM reporting," which the JPO ruled out because it wanted to maintain this reporting

- "engineering change proposals funded separately," which is simply a transfer of costs to a different contract that does not reduce overall cost to the governments.

Summary

In this chapter, we described the methodologies used in our assessment of EOQ, CRI, and other sources of savings; details of our approaches are not described in this unrestricted portion of the report to protect the contractor's proprietary data underlying the model. In the next chapter, we integrate these methodological pieces to generate estimates for overall BB savings.

CHAPTER FOUR

BB Savings Estimates

Drawing upon our approaches to assessing different sources of BB savings, we constructed a cost model for the F-35 BB contract. The model begins with the annual contracting baseline, then applies savings from each of the savings categories we identified to arrive at a BB contract cost, from which we can derive the percentage of savings relative to a series of annual contracts. In this chapter, we briefly describe the model, including the parameters that vary, then present overall BB savings results. We put these results in context by comparing them with estimated savings on previous MYP contracts for fighter aircraft. Finally, we discuss potential areas of risk that could reduce these savings and should be considered as part of a full business case analysis.

Monte Carlo Model

Our Monte Carlo model allows several parameters to vary, including the following:

- **annual contracting baseline.** We can use the model to compute BB savings relative to different annual contracting cost baselines, such as the RAND and JPO estimates presented in Chapter Two. In this chapter, all results presented are relative to the RAND annual contracting baseline, unless otherwise stated. In the Monte Carlo analysis, we assign a baseline cost distribution ranging from the RAND low to the RAND high estimates and centered on the RAND mid estimate (see Chapter Two for these estimates). The percentage savings resulting from the Monte Carlo analysis can then be normalized to any particular baseline.

- **government and contractor funding for CRIs.** As a default, the model assumes $100 million of government-funded CRIs in each year of the BB and includes additional contractor-funded CRI investment up to the annual cap on total CRI investment, which is set at 2 percent of the total contract cost. In the Monte Carlo analysis, the cap on CRI investment is varied in a triangular distribution centered at the 2-percent cap and ranging from zero contractor investment (government investment only) to a cap of 4 percent. As discussed in Chapter Three, we apply government-funded CRI savings to both the baseline and BB costs and contractor-funded CRI savings to the BB cost.
- **EOQ funding.** The default is that 4 percent of the contract value is available for government EOQ funding, but this cap can be varied to account for such effects as greater efficiency in EOQ activities or rejection of EOQ activities due to obsolescence risk. We also considered the possibility that additional savings might accrue to the government through contractor funding of EOQ. However, given the estimated supplier returns on EOQ funding, this funding mechanism was found to offer very little benefit.[1] In the Monte Carlo analysis, EOQ funding available is varied between 2 and 8 percent with a triangular distribution centered on 4 percent. This is not meant to represent uncertainty in availability of government funding so much as uncertainty in the return on investment in terms of savings that will be achieved with that investment.
- **degree of savings extrapolation through supply base.** Our default assumption is that suppliers for whom we do not have

---

[1] The reason that contractor funding of EOQ did not offer much benefit in our analysis is because the assumed required rate of return for contractors (13 percent) was sufficiently large relative to estimated supplier returns on EOQ funding that it was to the advantage of the governments to fund EOQ with their own money. However, as government EOQ funding becomes increasingly constrained, contractor funding of EOQ begins to generate savings beyond what government funding of EOQ generates. This is seen in the next chapter in the context of the hybrid BB analysis.

ROM estimates behave like suppliers for whom we do have such estimates. We can vary this assumption such that nonresponding suppliers offer no, or reduced, savings. For the results shown here, we did not vary this parameter.

To generate a savings range, we assigned parameter distributions as described and built a Monte Carlo simulation that was run 5,000 times.

BB Savings Results

Figure 4.1 shows the simulation results as a frequency distribution of estimated savings. Because the F-35 air vehicle and engine are contracted separately, we show the results for each.[2] The medians of the distributions are about $1.8 billion for the air vehicle, $280 million

Figure 4.1
Distribution of BB Savings Estimates for F-35 Air Vehicle and Engine

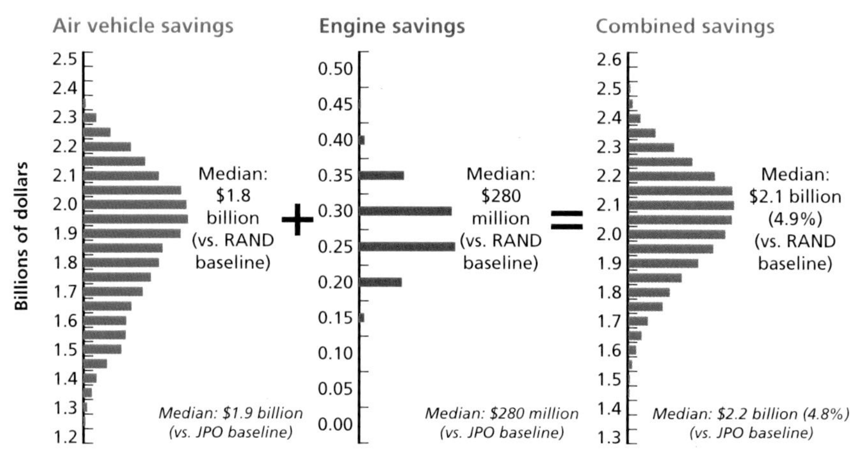

RAND *RR2063-4.1*

[2] Note that, while the air vehicle and engine are contracted separately and we report estimates for each of them, they were not treated completely independently in our analysis. The systems were linked in the analysis through the provision of EOQ funding, in the manner described in Chapter Three.

for the engine, and about $2.1 billion overall (or 4.9 percent of annual contracting cost). As noted earlier in this chapter, we focus on savings relative to the RAND annual contracting baseline in this report. However, we mention that the analogous savings point estimate using the JPO annual contracting baseline is $2.2 billion—$1.9 billion for the air vehicle and $280 million for the engine—or 4.8 percent of the annual contracting cost.

Figure 4.2 shows the range of estimates for total savings using a box-and-whisker plot. On the right of the figure, the box covers the range from the 25th percentile ($1.9 billion) of the savings distribution at the bottom edge to the 75th percentile ($2.2 billion) at the top edge. The median of the distribution ($2.1 billion) is indicated by the line through the box. The whiskers indicate the tenth ($1.8 billion) and 90th ($2.3 billion) percentiles of the distribution.

As a reminder, when we refer to BB savings, we mean only those savings that arise specifically from having a BB contract in place and that would not be available under annual contracting. Thus, BB sav-

Figure 4.2
Range on Overall BB Savings Estimate

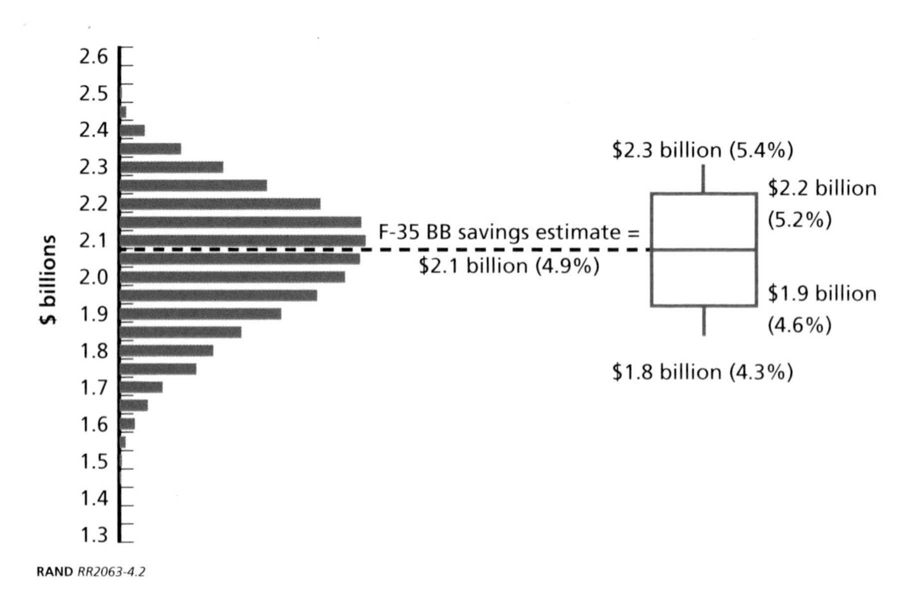

ings do not include reductions in cost from natural learning and rate effects or from government-funded CRIs, neither of which require a BB contract. BB savings do include savings from EOQ activities, management challenge and fee reduction (much of which we estimate using a contractor-funded CRI model), and a host of other smaller categories of savings described in Chapter Three. A breakdown of our overall BB savings into these categories is presented in Table 4.1 for our median estimate. In addition, our BB savings estimates include only recurring flyaway costs of the aircraft; our estimates do include savings in other cost elements, such as initial spares, ancillary equipment, and nonrecurring tooling. Finally, note that the estimates are not discounted for the risks associated with achieving these savings (discussed further later).

Allocation of BB Savings to Countries

The previous discussion presents the *overall* estimated BB savings. What is the allocation of BB savings to individual countries? The BB cost model calculates per-unit savings by variant and by lot, so one approach to allocating savings is simply to assign the savings calculated for each individual aircraft to the country that purchases that aircraft. Tables 4.2 and 4.3 summarize the results of allocating country-by-country BB savings this way, for the RAND and JPO annual contracting baselines, respectively. This allocation method tends to reward countries for committing to purchasing aircraft in later lots, largely because of how savings from CRI investments phase in over time. This

Table 4.1
Breakdown of BB Savings Median Estimate into Savings Categories

Savings Category	Savings ($ millions)	Total Savings (%)[a]
EOQ	790	38
Contractor-funded CRI	780	38
Administrative and other	510	25

[a] Percentages do not sum to 100% due to rounding.

Table 4.2
Allocation of BB Savings by Country Using RAND Annual Contracting Baseline

Program	Lot 12 Quantity	Lot 13 Quantity	Lot 14 Quantity	BB Savings ($ millions)
AU—F-35A	15	15	15	237
NL—F-35A	8	8	8	126
TR—F-35A	8	8	8	126
UK—F-35B	3	6	8	102
FMS-1—F-35A	22	22	5	235
CA—F-35A	0	4	9	81
DK—F-35A	0	4	6	61
IT—F-35A	2	2	4	45
IT—F-35B	1	3	5	56
NO—F-35A	6	6	6	95
US—F-35A	44	48	48	744
US—F-35B	20	20	20	334
US—F-35C	6	12	18	237
FMS-2—F-35A	6	6	6	95

Table 4.3
Allocation of BB Savings by Country Using JPO Annual Contracting Baseline

Program	Lot 12 Quantity	Lot 13 Quantity	Lot 14 Quantity	BB Savings ($ millions)
AU—F-35A	15	15	15	252
NL—F-35A	8	8	8	134
TR—F-35A	8	8	8	134
UK—F-35B	3	6	8	108
FMS-1—F-35A	22	22	5	250
CA—F-35A	0	4	9	86
DK—F-35A	0	4	6	65
IT—F-35A	2	2	4	48
IT—F-35B	1	3	5	59
NO—F-35A	6	6	6	101
US—F-35A	44	48	48	790
US—F-35B	20	20	20	355
US—F-35C	6	12	18	252
FMS-2—F-35A	6	6	6	101

effect can be seen, for example, by comparing the F-35A savings allocated to Australia versus those allocated to a subset of FMS countries (FMS-1). Australia buys fewer aircraft (45 vs. 49) but receives greater savings because they commit to so many of their aircraft in lot 14. In some sense, this allocation makes sense: It is the willingness to commit to purchases in out-years that drives BB savings. However, such an allocation approach can create perverse incentives, such that any individual country is better off delaying its purchases to the out-years, despite the fact that the success of the BB and the program itself depends on relatively constant quantities in each of the three BB years. We therefore present this allocation only as an example: We do not believe there is an objectively optimal method for allocating savings.

Comparison with Previous MYP Contracts

How do the estimated savings for an F-35 BB contract compare with estimated savings from historical MYP and BB fighter programs? Every historical aircraft MYP contract has been unique in terms of both the content and the approach taken. Thus, caution should be exercised in making comparisons across programs, including comparisons to the potential F-35 BB contracts. As reported in Appendix B, we conducted an extensive regression analysis of multiple factors from historical MYPs, such as program length or program maturity, which may be thought to affect the scale of savings for MYPs. This regression analysis showed no statistically significant correlation between any factor we tested and the scale of historical savings estimates (though it bears noting that the sample size of multiyear contracts is small). This finding applies to the three historical fighter aircraft MYPs discussed in Appendix B—F-16 MYP I, F/A-18E/F MYP I, and F-22 MYP—which may appear to be the best candidates for comparison to the F-35.[3] In addition, these three historical fighter MYPs possessed a variety of significantly different program attributes beyond those factors

---

[3] As noted earlier in this report, there are no previous examples of BB contracts for fighter aircraft.

tested in the regression analysis. It is not possible to determine whether any of these differing attributes have any effect on MYP cost savings estimates. Nonetheless, along with the factors already discussed, it is useful to keep these differences in mind when making cross-program comparisons of historical fighter MYPs. Some of the specifics of these variations and differences are discussed below.

Figure 4.3 compares our results with the estimated savings for F-16, F/A-18 E/F, and F-22 multiyear contracts.[4] In all cases, we use the first multiyear contract for comparison. Note that these are not validated, realized savings; they are pre-multiyear estimates as reported in justification packages. Also, the F-16 and F/A-18 E/F MYP contracts covered only the air vehicle, not the engines. However, this does not significantly affect the comparison: Our calculated savings percentages for the F-35 air vehicle and the full aircraft are quite similar.

Note that the historical programs vary in their treatment of government-funded CRIs: The F-16 and F-22 MYP contracts did not include government-funded CRIs among BB savings, but the F/A-18 E/F MYP contract did. We therefore present our F-35 BB savings estimate both with and without government-funded CRIs included among BB savings. Our savings estimate of 4.9 percent for the F-35 BB is lower than the 6.1 percent average multiyear savings estimate of F-16 and F-22, the two historical fighter programs that did not have government-funded CRIs. When savings arising from government-funded CRIs are added, our resulting F-35 savings estimate of 6.1 percent is lower than the 7.4 percent savings estimate for F/A-18E/F, which similarly includes savings from government-funded CRIs. Thus, in both cases, the F-35 savings estimate is roughly comparable, if lower, than those of historical fighter programs.

Given the brief overviews of the differences between the three other historical fighter MYPs and the F-35 BB, caution should be used when attempting to compare multiyear savings. Quantitative analysis is unable to demonstrate which factors, if any, are most important in achieving MYP savings. Furthermore, all of these numbers are merely estimates, and we know from the historical analysis that it is nearly

---

[4] There are no previous examples of BB contracts for fighter aircraft.

Figure 4.3
F-35 BB Overall Savings Estimate vs. First Multiyear Savings Estimates for Other Fighter Aircraft

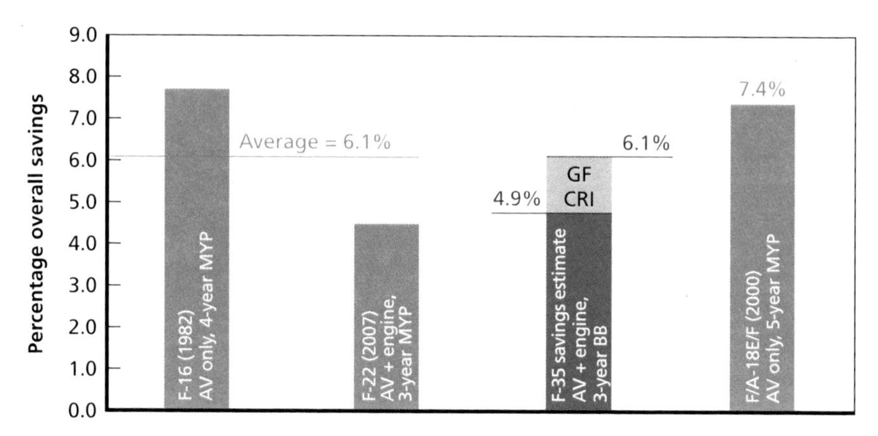

NOTE: The F-16 and F-22 multiyear contracts did not include government-funded (GF) CRIs, but the F/A-18 E/F multiyear contract did. Other differences among programs are indicated in the plot, including contract starting year, length of contract, and whether both air vehicle (AV) and engine were included.
RAND *RR2063-4.3*

impossible to determine the actual savings achieved by any historical MYP once it was executed. All we can say for certain is that RAND's estimates for the F-35 BB are close to those of the F-16 MYP I, F/A-18E/F MYP I, and F-22 MYP. This information indicates that the RAND savings estimates for the F-35 BB are within reasonable expectations and parameters, given historical experience.

Potential Risks Associated with BB Contracts: Areas for Further Analysis

There are several risks that must be managed to achieve BB savings, including availability of early EOQ funding, configuration changes, and aircraft quantity reductions. While a full risk analysis is beyond the scope of this research, we note a handful of risks in this section. These risks must be considered as part of a full cost-benefit analysis on whether to proceed with a BB contract, and should be actively managed if such a contract is ultimately pursued for the F-35 program. Most of the risks discussed here are general concerns about BB or MYP contracting approaches, but we begin with one risk that is specific to our data and analysis: availability of early EOQ funding.

The government's intent for EOQ funding, at the time of this analysis, was to provide half of the EOQ funding in FY 2017 and half in FY 2018. Suppliers who provided a ROM savings estimate also provided a schedule for when they would require EOQ funding. In our main analysis, we used only the total amounts when distributing EOQ funding; we did not account for the timing but assumed that small adjustments in timing should be possible if needed, particularly because the suppliers were not given constraints on timing when preparing their initial responses. However, we did roll up the supplier data to examine the timing of the funding requests. As it turns out, and as shown in Table 4.4, the requested EOQ funding is heavily weighted toward FY 2017. It is unclear whether the contractors will be able to smooth the EOQ funding profile without sacrificing savings. For some countries, including the United States, the availability of FY 2017 EOQ funding poses a risk, as requests for EOQ funding in the defense budget often need to happen many months in advance. For this reason, Chapter Five examines an alternative ("hybrid") BB approach that assumes limited government EOQ funding in FY 2017.

We also note that because of EOQ, other AP, or other CRI funding by the government, multiyear contracts will typically have more front-loaded funding in their earlier years. This increased expense can come at the opportunity cost to other programs within a service's port-

Table 4.4
Requested EOQ Funding Distribution by FY

FY	EOQ Funding Requested (TY $ billions)
2017	1.15
2018	0.66
2019	0.02

folio that could have been further funded during these "bow-wave" fiscal years.

The literature on BB savings and the DoD primarily stress the loss of flexibility in funding; requirements; and schedule for the military services, DoD, and Congress. Although multiyear contract authorization technically requires that programs demonstrate stability in these areas before award, unforeseen circumstances with design or quantity requirements, as well as budget environments, can force the government to consider trade-offs between maintaining the program's current trajectory, engaging in potentially high-cost renegotiations on certain contract clauses, or incurring termination or cancellation costs. If the latter two appear too costly for the multiyear contract, other, annually contracted programs within the portfolio may have to sustain funding or schedule impacts.[5]

Another risk to achieving BB savings is the potential for configuration changes that occur during the BB period. A BB contract could induce additional costs for configuration changes, primarily in two ways:

- EOQ funding and purchases could result in additional funding being sunk on components that are no longer needed.
- Production acceleration (also tied to EOQ funding) could result in a greater number of units requiring rework.

It is possible that this risk may be reduced by providing EOQ funding only to those components deemed to have the lowest likelihood of changes. This, however, could reduce potential savings.

Yet another risk to achieving full BB savings is related to aircraft quantity reductions that may occur during the BB period. This could erode savings for the following three reasons:

- Existing suppliers may be less motivated to provide the same quantity discount on parts and material.
- New suppliers may have less incentive to compete for workshare.

---

[5] O'Rourke and Schwartz, 2015.

- Fewer CRIs would be implemented if there were fewer aircraft over which to recoup investments.

Note that it is also possible that some savings may be at risk if contractors and their suppliers have the *perception* that the quantities specified may be reduced, even if such changes have not been announced.

We reiterate that the JPO must analyze these and potentially other risks before making a decision to proceed with a BB contract for F-35. The savings estimates presented in this report are only one part of the larger tradespace to be considered.

Summary

In this chapter, we described our F-35 BB cost model and presented the savings estimates generated by this model. The total estimated savings of $2.1 billion, or 4.9 percent of annual contracting costs, are roughly comparable to the estimated savings on previous MYP contracts for fighter aircraft. However, a BB contract carries some potential risks that should be considered as part of a full business case analysis. While a detailed risk assessment is beyond the scope of this research, we respond to one potential concern—the availability of EOQ funding in FY 2017—in Chapter Five by examining a hybrid BB approach.

CHAPTER FIVE

Hybrid BB: An Analytic Excursion

Thus far, we have analyzed a three-year F-35 BB contract that includes all partner countries and covers lots 12–14, with EOQ funding beginning in FY 2017. However, as noted in Chapter Four, the availability of EOQ funding in FY 2017 is uncertain at the time of this writing.[1] As an analytic excursion, the F-35 JPO asked PAF to consider alternative "hybrid" BB constructs that commit only a subset of partner countries to the BB contract for lots 12–14 and gives the remaining countries the option of joining after the first year. This chapter describes these hybrid BB constructs, summarizes our analysis methodology, and presents our estimate of savings for these constructs. As an additional comparison, we show savings for a case in which all countries engage in the BB contract, but no EOQ funding is available until FY 2018.

Overview of Hybrid BB Constructs

The hybrid BB approach is depicted in Figure 5.1. We divide the 471 aircraft included in lots 12–14 into two groups according to whether they are part of the initial BB contract or the later option. Group 1 (green) comprises 244 aircraft procured under the initial BB contract. These include all lot 12 aircraft for all partner countries (141 aircraft), plus lots 13 and 14 aircraft for the "participant nations"—United Kingdom,

[1] U.S. participation in a lot 12–14 BB contract was contingent upon Congress providing EOQ funds in FY 2017. The FY 2017 President's Budget only provided EOQ funding in FY 2018 for the U.S. military services to enter a BB contract in lot 13.

Figure 5.1
Outline of the Hybrid BB Construct

		Lot 12	Lot 13	Lot 14	
Participant nations	AU—F-35A	15	15	15	Group 1 224 aircraft
	NL—F-35A	8	8	8	
	TR—F-35A	8	8	8	
	UK—F-35B	3	6	8	
	FMS-1—F-35A	22	22	5	
Option nations	CA—F-35A	0	4	9	Group 2 227 aircraft
	DK—F-35A	0	4	6	
	IT—F-35A	2	2	4	
	IT—F-35B	1	3	5	
	NO—F-35A	6	6	6	
	US—F-35A	44	48	48	
	US—F-35B	20	20	20	
	US—F-35C	6	12	18	
	FMS-2—F-35A	6	6	6	
	Total:	**141**	**164**	**166**	**471**

RAND *RR2063-5.1*

Australia, the Netherlands, Turkey, and FMS-1 (an additional 103 aircraft). Group 2 (red) comprises the 227 aircraft that are not included in the initial BB contract. However, we assume there is an option, to be exercised by the first quarter of FY 2018, that allows the "option nations"—the United States, Canada, Denmark, Italy, Norway, and FMS-2—to add their lots 13–14 aircraft to the BB contract.

We examined two variations on this concept, shown in Table 5.1: one in which each country provides 4 percent EOQ funding for only the lots it procures as part of the BB ("Hybrid 1") and another in which all countries have 4 percent EOQ funding for all three lots, but the contractors are incentivized to provide half of the funding in FY 2017 for the option nations with a 20-percent return on investment one year later ("Hybrid 2"). Hybrid 1 would result in a smaller amount of EOQ funding, especially in FY 2017, compared with the original BB construct because the option

Table 5.1
Hybrid 1 and Hybrid 2 EOQ Funding

	Hybrid 1 EOQ Funding		Hybrid 2 EOQ Funding	
Participant	**FY 2017**	**FY 2018**	**FY 2017**	**FY 2018**
Participant nations	2% of lots 12, 13, 14	2% of lots 12, 13, 14	2% of lots 12, 13, 14	2% of lots 12, 13, 14
Option nations		4% of lots 13, 14 *only*		2% of lots 12, 13, 14
Contractors (for option nation aircraft)			2% of lots 12, 13, 14[a]	

[a] To be reimbursed by option nations with 20-percent return on investment after one year.

nations would contribute EOQ funding only for lots 13 and 14, and only in FY 2018. Hybrid 2 would result in the same amount of EOQ funding in FYs 2017 and 2018 as the original BB concept, but at an additional cost to incentivize the contractors to invest. In both variations, if the option is not exercised, then the option nations will contract annually for their lot 13 and 14 aircraft, and no EOQ funding will be available from them.

Methodology for Assessing Hybrid BB Savings

The first step in assessing the savings of the hybrid BB constructs is to estimate the annual contracting cost for this production profile. Unless otherwise stated, we use the RAND annual contracting baseline computed in Chapter Two. In the next subsection, we discuss how we collected data to support our hybrid BB savings assessments.

Contractor Questionnaire

To assess hybrid BB savings, we needed additional data from the prime contractors and major suppliers. However, given the limited resources and schedule available to RAND and the F-35 contractors for this analysis, we used a short questionnaire that, to the greatest extent possible, drew upon the data already provided to us to support the analysis of the original BB construct. More specifically, the questionnaire employed multiple-choice questions that indicated how respondents' original BB EOQ and non-EOQ savings estimates ought to be adjusted to account for the hybrid BB approach. These options included:

- preserving the percentage savings relative to the original BB construct
- scaling the percentage savings for Group 1 and Group 2 according to the size of these groups, relative to the size of the original BB
- setting the percentage savings to zero.

In addition, multiple-choice questions were provided to the contractors to identify how their original EOQ funding requests should be transformed into Group 1 and Group 2 EOQ funding requests for the Hybrid 1 BB.[2] For all multiple-choice questions, we also allowed the contractors to provide their own estimate rather than select one of the options we provided. Finally, we asked the contractors to provide justifications for their responses to the multiple-choice questions.

Hybrid BB Cost Model

The modeling approach for estimating hybrid BB savings was largely the same as that used for the original BB. Savings were evaluated in the same three categories—EOQ, CRI, and administrative and other—but with the following adjustments.

EOQ Savings

In the Hybrid 1 analysis, EOQ savings were calculated independently for Group 1 and Group 2 aircraft. For each group, the savings ROMs provided by suppliers in the hybrid questionnaires were ranked and funded in order of their dollars of savings returned per dollar of EOQ funding provided. The savings for each group were established independently in this way and summed to determine the total EOQ savings. There was one significant difference in the way that EOQ funding was constrained. Recall that in the original BB analysis, we limited EOQ funding to 4 percent of the total annual contract value, and

[2] As noted earlier, the EOQ profile for the Hybrid 2 BB is the same as for the original BB.

assumed that any inequality between FY 2017 and FY 2018 funding could be managed during contract negotiations without sacrificing savings. In the hybrid analysis, because only Group 1 aircraft are entitled to FY 2017 EOQ and because only participant nations provide EOQ, we had to apply separate constraints on EOQ funding. For Group 1, FY 2017 EOQ funding was capped at 2 percent of the annual contract value for participant nations only. For Group 2, FY 2018 EOQ funding for option nations was capped at 4 percent of the annual contract value for lots 13 and 14 only. These constraints and the supplier questionnaire responses, which generally offered lower EOQ returns for the hybrid BB than for the original BB, both served to reduce the EOQ savings available under the hybrid BB.

For Hybrid 2, the analysis is simpler. The contractors provide the missing FY 2017 EOQ funding, so the total EOQ savings is the same as it was in the original BB analysis. However, in this case, the governments do not see the whole savings, because they must pay the contractors 20 percent of the EOQ investment as an incentive.

CRI Savings

Government CRIs, which are independent of contract type and length, are unchanged in the hybrid analyses. However, contractor-funded CRIs (which we analyze as an estimate of the management challenge that can potentially be taken on by contractors) do need to be evaluated differently. Because the Group 2 aircraft are an option at the beginning of the hybrid period, the contractors cannot reasonably be expected to factor them into the expected returns from CRI investment. Therefore, for both hybrid BB constructs, contractor investment in the first year will be on the basis of Group 1 aircraft alone to justify the investment. As it turns out, the quantity and timing of Group 1 aircraft are in fact sufficient to justify investment to the 2-percent ceiling previously described—and, in the end, CRI savings are exactly the same in the hybrid constructs as they were in the original BB construct.

Administrative and Other Savings

In the hybrid questionnaire responses, many contractors and suppliers provided estimates of savings in this category independently for Groups 1 and 2 (some responded that the savings would be the same as in the original BB construct, and others responded that there would be no savings in the hybrid BB). As with the original BB, we used the contractor-provided ROM estimates, adjusted to the appropriate baseline annual costs, and extrapolated to account for nonresponding suppliers. We carried out this analysis independently for Group 1 and Group 2 aircraft, and summed them to determine a total savings.

Evaluating the hybrid BB required building separate savings models for Group 1 and Group 2 aircraft, while manually accounting for the effects of early CRI and EOQ investment on Group 2 aircraft costs. Owing to the changes to the modeling structure and time constraints, we were unable to implement Monte Carlo analysis of the hybrid BB. Thus, the results shown in this section are point estimates, using the RAND baseline estimate and the assumptions described in Chapters Two and Three. For the original BB analysis, the median savings from the Monte Carlo analysis was very close to the point estimate, so we expect the results to be comparable for the hybrid BB, as well.

Results and Discussion

Overall Savings Results

Table 5.2 compares the estimated savings for the original and hybrid BB constructs. For additional context, we add a fourth option that assumes a BB contract with all countries participating, but no EOQ funding in FY 2017. Government EOQ funding equal to 4 percent of the lots 13–14 cost is assumed to be available beginning in FY 2018.[3] Table 5.3 presents the analogous savings figures if the JPO annual contracting baseline is

[3] EOQ savings are estimated using the average EOQ savings return on EOQ funding for Group 2 aircraft in the Hybrid 1 BB construct. This is a reasonable assumption because Group 2 aircraft in the Hybrid 1 BB construct are eligible for EOQ funding beginning in FY 2018, just like aircraft in lots 13–14 in the "BB, No FY 2017 EOQ" approach. However, because the number of aircraft in Group 2 of the hybrid BB approach is less than in lots 13–14 of the "BB, No FY 2017 EOQ" approach (227 vs. 330), it could be argued that this assumption leads to a conservative estimate of EOQ savings (i.e., a greater number of aircraft, through enhanced economies of scale, might lead to better returns on EOQ funding). Non-EOQ savings are the same as in the original BB construct.

Table 5.2
Comparison of BB Savings for BB Approaches Using RAND Annual Contracting Baseline

Approach	Savings ($ billions)	Percentage of Annual Contracting Cost	Percentage of Original BB Savings
Original BB	2.1	4.9	—
Hybrid 1 BB	1.6	3.7	77
Hybrid 2 BB	1.8	4.1	89
BB, no FY 2017 EOQ	1.6	3.9	79

Table 5.3
Comparison of BB Savings for BB Approaches Using JPO Annual Contracting Baseline

Approach	Savings ($ billions)	Percentage of Annual Contracting Cost	Percentage of Original BB Savings
Original BB	2.2	4.8	—
Hybrid 1 BB	1.6	3.7	76
Hybrid 2 BB	1.9	4.2	87
BB, no FY 2017 EOQ	1.7	3.9	78

used instead of the RAND annual contracting baseline; however, the following discussion refers to the savings figures in Table 5.2.

As Table 5.2 illustrates, the Hybrid 1 BB construct achieves $1.6 billion in savings compared with the annual contracting baseline. This corresponds to 77 percent of the $2.1 billion of BB savings estimated for the original BB construct. Several factors drive this reduction in savings. One is that EOQ savings are approximately 44 percent lower for the Hybrid 1 construct than for the original construct. This results from two factors:

- There is less EOQ funding in the Hybrid 1 BB than in the original BB because 85 of the lot 12 aircraft in Group 1 are ineligible for EOQ funding.
- By necessarily treating Group 1 and Group 2 EOQ as separate procurements in the Hybrid 1 BB, there are reduced economies of scale compared with bundling them together, as in the original BB. Thus, there is an expectation that each dollar of EOQ fund-

ing should return less savings in the Hybrid 1 BB than in the original BB.

Savings in the "administrative and other" category detailed in Chapter Three are reduced by approximately 17 percent in the Hybrid 1 BB relative to the original BB construct. As in the case of EOQ, reduced economies of scale in the Hybrid 1 BB are a significant driver for this reduction in savings. In addition, contractors cited the complexity of bidding, proposing, negotiating, and managing the Hybrid 1 BB construct as a reason for reducing savings.

Perhaps unexpectedly, the Hybrid 1 BB construct achieves the same savings from contractor-funded CRI investments as the original BB construct. In assessing contractor-funded CRI savings, we applied the principle described in Chapters Three and Four of determining how much the contractors would be willing to invest in CRIs, given the quantities of aircraft committed at that point. While far fewer aircraft are committed initially in the Hybrid 1 BB than in the original BB (244 aircraft, as opposed to 471), the reduced quantities are sufficiently large and the anticipated CRI returns are sufficiently high that contractors would maintain the level of CRI investment assumed in the original BB. Thus, available savings from contractor-funded CRIs are the same in the Hybrid 1 BB. However, because the Hybrid 1 BB presents a greater risk to the contractors of not recouping their investments owing to fewer committed aircraft, it may be reasonable to expect that negotiating these savings would be more challenging in the case of the Hybrid 1 BB.

As Table 5.2 illustrates, the Hybrid 2 BB achieves $1.8 billion in savings compared with the annual contracting baseline. This corresponds to 87 percent of the $2.1 billion of BB savings estimated for the original BB construct. Given the similar structures of the Hybrid 1 and 2 BB constructs, their estimated CRI and other savings are identical. Thus, it is only EOQ savings that drives the difference in savings between these two approaches. As noted above, the EOQ funding profile for the Hybrid 2 BB is the same as for the original BB, so their EOQ savings are identical. However, because a portion of the EOQ funding is provided by contractors in the Hybrid 2 BB, its savings are decremented to account for the contractor return on EOQ investment, which we assume to be 20 percent. Thus, EOQ savings in the Hybrid 2 BB construct are approximately 80 percent of those in the original BB construct.

As an excursion to our Hybrid 2 BB savings analysis, we assessed the available savings if the contractors provided FY 2017 EOQ funding beyond 2 percent of the contract cost for option nations. EOQ savings generated as a function of total EOQ provided is illustrated in Figure 5.2. The dashed vertical line indicates the point at which the available EOQ funding from participant nations runs out, and where this vertical line crosses the purple line corresponds to our $1.8 billion savings estimate. Savings increase as the contractors provide additional EOQ funding, as long as the newly funded suppliers provide enough savings for the contractors to recoup their EOQ investment at a 20-percent rate of return. The maximum additional savings available

Figure 5.2
EOQ Savings vs. EOQ Funding for the Hybrid 2 BB

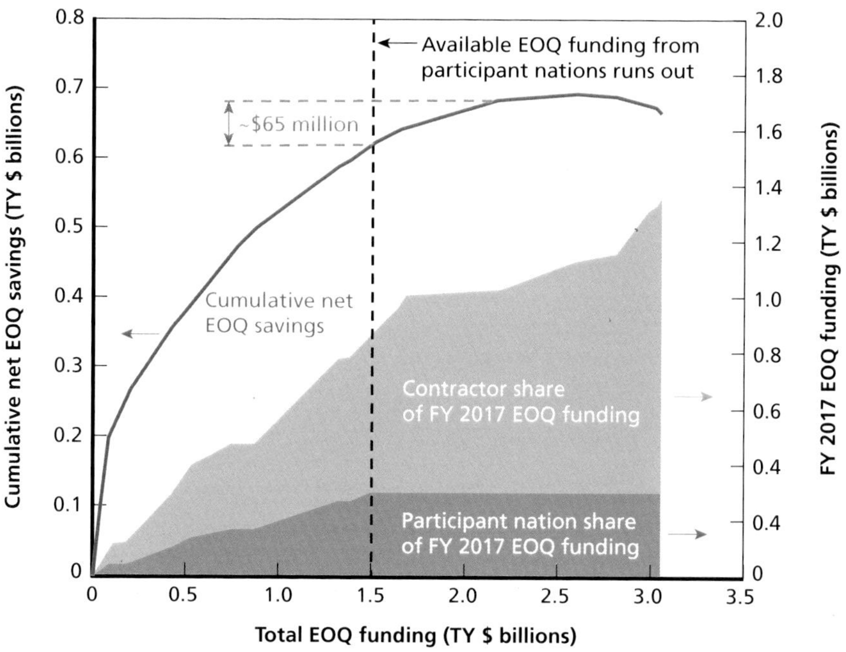

RAND *RR2063-5.2*

by increasing contractor EOQ investment is approximately $65 million, as shown in the figure.

As noted, we also considered a fourth construct that assumes a BB contract involving all partners and all three lots but no EOQ funding in FY 2017. This approach reflects the possibility that the United States and some partner countries will be unable to make EOQ funding available in FY 2017, but it avoids the complexity of negotiating and managing contracts associated with the hybrid BB approach. As shown in Table 5.2, this approach would save $1.6 billion compared with the annual contracting baseline, or slightly more savings than the Hybrid 1 BB. It should be noted that the "no FY 2017 EOQ" approach would require BB authorization from the United States and other option nation governments a year earlier than for the hybrid BBs, where the decision of whether to exercise the Group 2 option can be made a year later than the initial contract.

Tables 5.4 and 5.5 compare the estimated savings for the original and the alternative BB constructs considered in this chapter when sav-

Table 5.4
Comparison of BB and Government-Funded CRI Savings for Different BB Approaches Using RAND Annual Contracting Baseline

Approach	Savings ($ billions)	Percentage of Annual Contracting Cost	Percentage of Original BB Savings
Original BB	2.6	6.1	—
Hybrid 1 BB	2.1	4.9	81
Hybrid 2 BB	2.3	5.4	90
BB, no FY 2017 EOQ	2.2	5.1	84

Table 5.5
Comparison of BB and Government-Funded CRI Savings for Different BB Approaches Using JPO Annual Contracting Baseline

Approach	Savings ($ billions)	Percentage of Annual Contracting Cost	Percentage of Original BB Savings
Original BB	2.7	6.1	—
Hybrid 1 BB	2.2	5.0	81
Hybrid 2 BB	2.5	5.5	90
BB, no FY 2017 EOQ	2.3	5.2	83

ings from government-funded CRIs are included. Table 5.4 employs the RAND annual contracting baseline; Table 5.5 employs the JPO one. We include these savings estimates so that they could be compared to those of previous multiyear contracts that included savings from government-funded CRIs.

Allocation of Hybrid Savings to Countries

As with the original BB savings, in this section we address the allocation of hybrid BB savings to individual countries. However, in the case of the hybrid BB constructs, there is no objectively "correct" way to allocate these savings to individual countries. There are many possible approaches, and the one ultimately chosen will involve a negotiation among stakeholder countries. Nevertheless, we propose the following reasonable approach:

- For participant nations, allocate savings equal to what would be achieved in the original BB since they are making the same commitment of aircraft and EOQ funding as in the original BB construct.
- For option nations, allocate the remaining savings in proportion to their aircraft costs.

Tables 5.6 and 5.7 illustrate the results of applying this allocation scheme to the Hybrid 1 and 2 BBs, respectively. As shown in Table 5.6, this allocation scheme naturally provides greater savings to countries that commit to the three-year BB up front than it does for countries that exercise the BB option later on.[4] Tables 5.8 and 5.9 represent the analogous results using the JPO annual contracting baseline instead of the RAND annual contracting baseline.

---

[4] Note that if the allocation scheme did not do this, no individual country would rationally commit to the three-year BB upfront, and the hybrid BB approach would therefore not be executable.

Table 5.6
Allocation of Hybrid 1 BB Savings by Country Using RAND Annual Contracting Baseline

Program	Commit to Three-Year BB	Lot 12 Quantity	Lot 13 Quantity	Lot 14 Quantity	Original BB Savings ($ millions)	Hybrid 1 BB Savings ($millions)	Hybrid 1 BB Savings/Original BB Savings (%)
AU—F-35A	Yes	15	15	15	237	237	100
NL—F-35A	Yes	8	8	8	126	126	100
TR—F-35A	Yes	8	8	8	126	126	100
UK—F-35B	Yes	3	6	8	102	102	100
FMS-1—F-35A	Yes	22	22	5	235	235	100
CA—F-35A	No	0	4	9	81	58	72
DK—F-35A	No	0	4	6	61	44	72
IT—F-35A	No	2	2	4	45	32	72
IT—F-35B	No	1	3	5	56	40	72
NO—F-35A	No	6	6	6	95	68	72
US—F-35A	No	44	48	48	744	534	72
US—F-35B	No	20	20	20	334	240	72
US—F-35C	No	6	12	18	237	170	72
FMS-2—F-35A	No	6	6	6	95	68	72

Table 5.7
Allocation of Hybrid 2 BB Savings by Country Using RAND Annual Contracting Baseline

Program	Commit to Three-Year BB	Lot 12 Quantity	Lot 13 Quantity	Lot 14 Quantity	Original BB Savings ($ millions)	Hybrid 2 BB Savings ($ millions)	Hybrid 2 BB Savings/Original BB Savings (%)
AU—F-35A	Yes	15	15	15	237	237	100
NL—F-35A	Yes	8	8	8	126	126	100
TR—F-35A	Yes	8	8	8	126	126	100
UK—F-35B	Yes	3	6	8	102	102	100
FMS-1—F-35A	Yes	22	22	5	235	235	100
CA—F-35A	No	0	4	9	81	69	85
DK—F-35A	No	0	4	6	61	52	85
IT—F-35A	No	2	2	4	45	38	85
IT—F-35B	No	1	3	5	56	48	85
NO—F-35A	No	6	6	6	95	81	85
US—F-35A	No	44	48	48	744	636	85
US—F-35B	No	20	20	20	334	286	85
US—F-35C	No	6	12	18	237	203	85
FMS-2—F-35A	No	6	6	6	95	81	85

Table 5.8
Allocation of Hybrid 1 BB Savings by Country Using JPO Annual Contracting Baseline

Program	Commit to Three-Year BB	Lot 12 Quantity	Lot 13 Quantity	Lot 14 Quantity	Original BB Savings ($ millions)	Hybrid 1 BB Savings ($ millions)	Hybrid 1 BB Savings/Original BB Savings (%)
AU—F-35A	Yes	15	15	15	252	252	100
NL—F-35A	Yes	8	8	8	134	134	100
TR—F-35A	Yes	8	8	8	134	134	100
UK—F-35B	Yes	3	6	8	108	108	100
FMS-1—F-35A	Yes	22	22	5	250	250	100
CA—F-35A	No	0	4	9	86	62	73
DK—F-35A	No	0	4	6	65	47	73
IT—F-35A	No	2	2	4	48	35	73
IT—F-35B	No	1	3	5	59	43	73
NO—F-35A	No	6	6	6	101	73	73
US—F-35A	No	44	48	48	790	573	73
US—F-35B	No	20	20	20	355	258	73
US—F-35C	No	6	12	18	252	183	73
FMS-2—F-35A	No	6	6	6	101	73	73

Table 5.9
Allocation of Hybrid 2 BB Savings by Country Using JPO Annual Contracting Baseline

Program	Commit to Three-Year BB	Lot 12 Quantity	Lot 13 Quantity	Lot 14 Quantity	Original BB Savings ($ millions)	Hybrid 2 BB Savings ($ millions)	Hybrid 2 BB Savings/Original BB Savings (%)
AU—F-35A	Yes	15	15	15	252	252	100
NL—F-35A	Yes	8	8	8	134	134	100
TR—F-35A	Yes	8	8	8	134	134	100
UK—F-35B	Yes	3	6	8	108	108	100
FMS-1—F-35A	Yes	22	22	5	250	250	100
CA—F-35A	No	0	4	9	86	73	85
DK—F-35A	No	0	4	6	65	56	85
IT—F-35A	No	2	2	4	48	41	85
IT—F-35B	No	1	3	5	59	50	85
NO—F-35A	No	6	6	6	101	86	85
US—F-35A	No	44	48	48	790	675	85
US—F-35B	No	20	20	20	355	303	85
US—F-35C	No	6	12	18	252	215	85
FMS-2—F-35A	No	6	6	6	101	86	85

Summary

In this chapter, we presented and analyzed two variations of the hybrid BB, an alternative approach being considered to address the likely unavailability of FY 2017 EOQ funding from the United States and some partner governments. We also presented a reasonable approach for allocating hybrid BB savings to countries. For the two hybrid BB constructs considered, we estimate the total hybrid BB savings to be $1.6 billion and $1.8 billion, or 3.7 percent and 4.1 percent of the cost of contracting annually for the aircraft. These hybrid BB savings represent approximately 80 percent and 90 percent of the savings available in the original BB construct.

As an alternative to the hybrid BB constructs, we presented and analyzed a simpler BB approach that commits all partner countries to a three-year BB but assumes no EOQ funding available in FY 2017. This approach achieves slightly more savings than the Hybrid 1 BB, but it requires BB authorization from the United States and some partner countries a year earlier than the hybrid BB.

CHAPTER SIX

Conclusions

This report summarizes an assessment of cost savings available to the F-35 program through a BB contract for lots 12–14. The overall median recurring flyaway savings were estimated to be $2.1 billion ($1.8 billion for the air vehicle and $280 million for the engine), which is equivalent to 4.9 percent of the cost of contracting annually for the aircraft over lots 12–14. When compared with initial MYP contracts for other fighter aircraft, our median F-35 BB savings estimate is comparable, if slightly lower: 6.1 percent vs. 6.5 percent when savings from government-funded CRIs are included. However, in dollar terms, the potential savings from an F-35 BB contract would dwarf savings from these and other MYP or BB contracts.

Owing to concerns about the availability of EOQ funding in FY 2017 at the time of this writing, we also examined a hybrid approach, in which a subset of countries would enter a BB contract for lots 12–14, with the remaining countries possibly entering the contract for the latter two lots. We estimate the hybrid BB savings to be approximately 3.7 to 4.1 percent of the cost of contracting annually for the aircraft. These savings represent approximately 80 to 90 percent of the savings available in the original BB construct. We also analyzed a simpler BB approach that commits all partner countries to a three-year BB contract but assumes no EOQ funding is available in FY 2017. This approach achieves about 80 percent of the savings available in the original BB, but requires BB authorization from the United States and some partner countries a year earlier than in the hybrid BB approach.

While our analysis estimates the available savings from a BB contract, the actual savings achieved would result from contract negotiations. An examination of previous weapon system programs (described in Appendix B) indicates that comprehensive and in-depth analysis of contractor cost structure, particularly on the lower tiers, beyond what has typically been done in the past, can substantially increase estimated program savings, particularly during the contract negotiation phase.

We reiterate that this analysis focuses on potential cost savings from an F-35 BB contract and does not include a formal risk assessment. Potential areas of risk may include the availability of early EOQ funding, configuration changes, and aircraft quantity reductions. The JPO should consider these and other risks as part of its decisionmaking process and actively manage them if a BB contract is pursued.

APPENDIX A

BB and MYP Contracting

This appendix provides an overview of the multiyear contracting mechanisms available to modern defense programs: both formal MYP contracts and BB contracts. It explores the legal requirements to enter into an MYP or BB contract and the process by which a program receives approval.

In general, the law requires that programs pursuing a MYP must demonstrate that the benefits of this contracting strategy, largely consisting of cost savings, must outweigh the risks associated with the additional commitments and constraints that the MYP places on the government. These risks include significant changes in the quantity or production rate required, or in the product design, which could create additional costs or leave the government with excess or inadequate systems. The legal requirements for BB contracting are far less formal and restrictive, but congressional approval is still required before entering into such a contract, and Congress has generally set similar expectations for BB as for MYP contracts.

Legal Status and Requirements for MYP Contracts

Until the passage of statute 10 U.S.C. 2306b as part of the National Defense Authorization Act (NDAA) of FY 1982, no formal mechanism existed for government agencies to enter contracts for more than a single year's procurement of a component or system. Despite this, Congress had started authorizing informal multiyear contracts in the 1950s on a case-by-case basis to achieve cost savings associated with bulk material purchases, guarantees of long-term business for contractors, and the ability of contractors to optimize workforce and production facilities based on this guarantee of business. These early multiyear contracts were estimated to have achieved an average dollar savings of 10 to 20 percent per program.[1]

To enable similar multiyear contract cost savings for a larger number of procurement programs, Congress codified additional multiyear contract regulations in the FY 1982 NDAA as recommended by Deputy Secretary of Defense Frank Carlucci in his Acquisition Improvement Plan initiatives. Federal law governing MYPs contains several requirements, including:[2]

- The contract must show "substantial savings" compared with the estimated program costs if the program was executed through multiple single-year contracts and the savings estimates associated with the multiyear contract must be "realistic."
- The item being contracted for is a complete end item, it has a "stable design," and technical risks associated with it are "not excessive."
- The need for the minimum total quantity and production/procurement rate of the item will remain "substantially unchanged" for the proposed duration of the contract.
- The head of the agency intends to continue to request funding to support the planned production rate over the intended course of the contract.[3]
- The contract must be a fixed-price contract, the unit cost of the item must be "realistic," and the program to procure it must not

---

[1] See GAO, 2008.

[2] The original language of 10 U.S.C. 2306b, 2014, as established by the 1982 NDAA has been amended several times since 1982, and the most recent requirements are reflected here. Additionally, MYP regulations slightly differ if a program's contract value is greater or less than $500 million. MYP regulations applicable to the F-35 program (greater than $500 million contract value) are discussed here.

[3] Approval of an MYP contract does not result in allocation of funding for that program for the duration of the contract, but rather indicates government intention to pursue the program for that length of time. Funding for the contracted program must be allocated in subsequent years' budgets.

have experienced recent significant cost growth (e.g., Nunn-McCurdy Act cost growth breaches).
- The contract should not preclude the ability of an agency to provide for competition in the production of the item or to terminate the contract if the supplier's cost, quality, or schedule performance becomes poor.
- The contract should contain cancellation provisions that consider both recurring and nonrecurring costs of the contractor associated with production of the items to be delivered, "to the extent that such provisions are necessary."
- The contract may be up to five years in duration.

Before entering an MYP, Congress must agree that the above requirements are met, and to obtain agreement, the agency responsible for the program is required to submit formal documentation to Congress. The required documentation and process for pursuing congressional MYP approval are discussed below.

One aspect of the regulations that makes congressional approval and implementation more challenging is that several of the key terms are not defined precisely (e.g., "stable design," "realistic" cost estimate) and are judged by a high bar that has shifted since the initial version of the legislation. One example of this is the need to prove "substantial savings." Until 1990, the MYP statute defined a required minimum savings of 10 percent, which can be challenging to achieve. To enable more programs to qualify for MYPs and increase cost savings across additional government procurement agencies, the requirement to demonstrate 10-percent savings was revised to its current form. While no minimum savings is currently required to obtain MYP authority, demonstrating 10-percent savings is still considered a goal when submitting MYP documentation.[4] An important note regarding proving substantial savings: Soliciting both annual and multiyear proposals from a contractor can help prove substantial savings, but the costs and time required to prepare separate proposals can be significant and not in the government's or program's best interest. As a result, the FAR states that

[4] See GAO, 2008.

the head of a contracting activity may pursue only a multiyear proposal if obtaining both proposals is prohibitive in cost or schedule.

Another key term lacking a precise definition is what constitutes "realism" in the cost estimates justifying MYP contracts. A *realistic* cost estimate is typically considered to be one based on costs from already executed procurements of the item being contracted for or based on analogous costs from similar programs. Finally, the term *stable design* is typically interpreted to mean that major engineering design changes are not expected during the contract and the design of the item has already been built and/or tested. The clear intent of the requirement is that systems should be reasonably mature before authorization of an MYP.[5]

Two other notable characteristics of MYP contracts are important to consider for the F-35 program: the ability to utilize EOQ funding and the need for a cancellation penalty. EOQ savings are viewed as a significant source of savings for MYPs, as discussed in Chapter Three. The process to obtain congressional approval of an MYP is discussed in detail in 10 U.S.C. 2306b and can take from slightly more than one year to almost three years depending on whether EOQ funding is required. Contract value, amount of EOQ requested, and cancellation ceiling level also affect the MYP approval process and timeline. The MYP approval requirements applicable to the F-35 program are discussed later.[6]

Generally, to obtain MYP approval, a DoD agency is expected to request approval of appropriations bill legislative language in tandem with its formal budget request submittal for the following fiscal year.[7]

---

[5] Historically, there appears to be wide latitude for different programs on this issue. Ideally, a stable design would be viewed as one that has completed the LRIP stage, and is ready to enter into full-rate production. This has not always been the case in the past, however. Examples where systems were still in the very early stages of production or were still involved in development and operational test and evaluation when MYPs were approved include F/A-18E/F MYP I and the V-22 MYP I. See Appendix B for further detail on specific programs.

[6] MYP approval requirements differ for programs with any of the following: a contract value of less than $500 million, EOQ funding request of less than $20 million, or cancellation ceilings below $100 million. The potential F-35 BB contracts are above all these thresholds.

[7] The final President's Budget submittal is due to Congress every January; to support this, agency requests are submitted to the Office of Management and Budget by November of the prior calendar year.

In support of this, agencies are also required by law to submit MYP certification findings to Congress no later than March 1 of the year when the MYP approval request is to be made. MYP certification findings, also known as "service justification packages," include the following:

- documentation that the request meets these MYP legal requirements
- detailed budget explanations (discussions of EOQ, cancellation penalties, etc.)
- a Non Advocate Cost Estimate validation/assessment from the Cost Assessment and Program Evaluation (CAPE) office.

CAPE's cost assessment period has no established duration, but some DoD agencies' internal guidance suggests that it can take four or five months to complete the assessment, meaning that the whole process for obtaining MYP authority can be expected to take approximately one year.[8] If congressional review of the MYP justification results in an approval, discussion of the MYP approval and funding allocation are added to both nonappropriations legislation and an appropriations bill, respectively. The program is authorized to begin negotiating the contract once the laws containing the appropriate language are signed by the president.

A final congressional notification is required 30 days before contract award that restates or updates the information reported in the initial MYP request. This 30-day congressional notification is required based on contract value, amount of EOQ requested, and/or cancellation ceiling level. If funding for the cancellation penalty is not budgeted, this submittal is where an agency justifies this decision to Congress, including a discussion of where the funding would be taken from in the event of a contract cancellation.[9]

---

[8] See Department of the Navy, Deputy Assistant Secretary of the Navy–Air, *DASN (AIR) Multiyear Procurement (MYP) Guidebook*, November 10, 2010.

[9] If the cancellation penalty is not funded, which is usually the case, the agency must demonstrate that unused authorized funding would be available during each FY sufficient to cover the negotiated ceiling cost of cancellation for that FY.

A request for EOQ funding prolongs this timeline because the necessary documentation and approvals must occur before final approval of the EOQ funding. Given the approximately one year long process for MYP/EOQ approval, documentation for first-year EOQ funding on a multiyear contract should be submitted at least two years before the proposed contract placement date. Based on some DoD agencies' internal MYP guidance[10] and discussions with program offices with MYP experience, if the MYP request and certification (as well as the EOQ request) are submitted less than two years before the proposed contract award, it is unlikely to be approved for first-year EOQ funding for that contract.

While obtaining MYP authority from Congress provides significant assurance that future funding will be available for a particular program, funding for an MYP is not preallocated and Congress may decide to cancel (or modify the duration or purchased quantity) at any time during the contract. In the event that the contract is terminated early, contractors that have invested in the MYP product line would lose any benefit from that investment. Cancellation penalties are compensation ceilings agreed upon in advance by the government and contractor during negotiations and are added to MYPs to prevent this risk being priced into the MYP by the contractor.[11] It is important to note, however, that while there is benefit to specifying a cancellation penalty in an MYP, there is no requirement for an agency to fund an MYP cancellation penalty. If not funded, the agency must notify Congress of this during the MYP approval process.

Legal Status and Requirements for BB Contracts

A second multiyear contracting approach is the BB contract. Unlike MYPs, BBs were not formally created by a statute or acquisition regulation. Instead, this form of "unofficial" multiyear contracting was established by the U.S. Navy and has only been used twice. The first

---

[10] Department of the Navy, 2010.

[11] MYP regulations do not specify a required percentage level or dollar amount for a cancellation penalty.

of these unofficial multiyear contracts—later termed "BBs"—did not meet formally established MYP requirements. This was the four-year *Virginia*-class (SSN 774) submarine multiyear contract authorized in Section 121 of the FY 1998 NDAA.[12]

Legislative language authorizing formal MYPs typically references MYP legislation, namely the limitations on the MYP authority and the requirements for additional documentation and justification before execution of the MYP. The BB discussed in the FY 1998 NDAA does not, instead authorizing a multiyear contract with no mention of existing MYP legislation or limitations, with two exceptions. First, the NDAA states that advance procurement materiel (EOQ) is authorized for the second, third, and fourth *Virginia*-class submarines. Second, it states that the government shall not be liable for a cancellation penalty in excess of whatever is appropriated for the first four submarines. Comparably open-ended language in the 2010 NDAA defense appropriations bill approved two BB contracts for up to 20 Littoral Combat Ships (LCSs), with up to ten ships on each contract.

While there were some similarities between the *Virginia*-class and LCS BBs, the legislation that authorized the latter contained no mention of EOQ or cancellation penalties. Since there are no BB regulations and the *Virginia* and LCS BBs were different, it is difficult to establish a definitive set of BB characteristics. That said, the term *BB* has come to define a multiyear contracting mechanism that:[13]

- does not need to meet legal criteria or require specific documentation to be approved

[12] The *Virginia*-class BB was for the first four submarines of the class and therefore might have had difficulty demonstrating it had a "stable design" or "realistic" cost estimates.

[13] Since only two BB contracts have been executed to date and the term "block" was coined informally, the term is used inconsistently in the literature. "Block" can be used either in reference to the BB contracting mechanism, or groups of components/systems of a common type and design that are purchased together. An example of confusing uses of the term "block" is the name for the four procurement groupings of *Virginia*-class submarines. The *Virginia*-class submarine program has entered four separate multiyear contracts (Blocks I–IV), but only Block I used a BB while the remaining blocks were formal MYPs. When used for anything other than the *Virginia*-class Block I contract or the LCS contracts, "block" does not refer to a BB contract.

- can be for more than five years of procurements
- requires separate congressional EOQ approval
- is not required to include cancellation penalties[14]
- does not necessarily require a fixed-price type contract.

Unlike MYPs, approval and establishment of BBs does not follow a formal process defined by statute and regulation. Both existing BB programs were initiated under different circumstances. While neither program followed a defined process to obtain approval for the BB, both included detailed interactions between the Navy and Congress before inclusion of BB language in legislation. The purpose of these interactions was to confirm that significant cost savings would be achieved through the use of a BB contract. In both cases, industrial base concerns (e.g., maintaining industrial base health via long-term guarantees of business) were likely an additional factor in BB approval. Based on these examples, future DoD programs that pursue a BB will likely be required to prove that significant cost savings and other benefits can be achieved by pursuing a multiyear contract for that end item, even though formal MYP statutory requirements do not apply.

---

[14] O'Rourke and Schwartz, 2015.

References

Bakhshi, V. Sagar, and Arthur J. Mendler, *Multiyear Cost Modeling*, Fort Lee, Va.: Army Procurement Research Office, 1985

Buzacott, J. A., "Economic Order Quantities with Inflation," *Operational Research Quarterly (1970-1977)*, Vol. 26 No. 3, 1975, pp. 553–558.

CFR—*See* Code of Federal Regulations.

Code of Federal Regulations, Title 48, Federal Acquisition Regulations System, Chapter One, Federal Acquisition Regulations, SubChapter C, Contracting Methods and Contract Types, Part 16, Types of Contracts, Subpart 16.4, Incentive Contracts, Section 16.403-1, Fixed-Price Incentive (Firm Target) Contracts, December 15, 1994.

Defense Acquisition Management Information Retrieval (DAMIR), *Selected Acquisition Report: F-35 Joint Strike Fighter Aircraft (F-35)*, Washington, D.C.: U.S. Department of Defense, December 2015.

Department of the Navy, Deputy Assistant Secretary of the Navy–Air, *DASN(AIR) Multiyear Procurement (MYP) Guidebook*, November 10, 2010. As of February 23, 2016:
http://www.secnav.navy.mil/rda/Policy/dasnairmypguidebookv20november102010.pdf

DoD—See U.S. Department of Defense.

General Services Administration, Federal Acquisition Regulation, Part 217, acquisition.gov website, January 19, 2017. As of March 31, 2016:
https://www.acquisition.gov/?q=browsefar

GAO—*See* U.S. Government Accountability Office.

Goyal, S. K., "Economic Order Quantity Under Conditions of Permissible Delay in Payments," *Journal of the Operational Research Society*, Vol. 36, No. 4, 1985, pp. 335–338.

Harris, Ford W., "How Many Parts to Make at Once," *Factory: The Magazine of Management*, Vol. 10, No. 2, 1913.

Joint Strike Fighter Program, "F-35 Lightning II Program Fact Sheet: Selected Acquisition Report 2015 Cost Data," March 24, 2016. As of January 3, 2017: http://www.jsf.mil/news/docs/20160324_Fact-Sheet.pdf

Kovacic, William E., "Commitment in Regulation: Defense Contracting and Extensions to Price Caps," *Journal of Regulatory Economics,* Vol. 3, 1991, pp. 219–240.

Lowenthal, Franklin, "Cost of Prediction Error in the Economic Order Quantity Formula," *Managerial and Decision Economics*, Vol. 3, No. 2, 1982, pp. 95–98. As of March 31, 2016:
http://www.jstor.org/stable/2487413

O'Rourke, Ronald and Moshe Schwartz, *Multiyear Procurement (MYP) and BB Contracting in Defense Acquisition: Background and Issues for Congress*, Washington, D.C.: Congressional Research Service, R-41909, March 4, 2015.

Rogerson, William P., "Economic Incentives and the Defense Procurement Process," *Journal of Economic Perspectives*, Vol. 8, No. 4, Fall 1994.

Smouse, Ronald, and Paul Tetrault, *Joint Strike Fighter Avionics Cost Improvement Study (Learn, Rate, Step-Down, and Other Considerations)*, F-35 Joint Program Office, November 2002.

Tirole, Jean, "Procurement and Renegotiation," *Journal of Political Economy*, Vol. 94, No. 2, 1986.

U.S. Code, Title 10, Armed Forces, Section 2306a, Cost or Pricing Data: Truth in Negotiations, 2011.

———, Title 10, Armed Forces, Section 2306b, Multiyear Contracts: Acquisition of Property, 2014.

———, Title 41, Public Contracts, Subtitle I, Federal Procurement Policy, Division C, Procurement, Chapter 35, Truthful Cost or Pricing Data, 2010.

U.S. Department of Defense, Defense Federal Acquisition Regulation Supplement (DFARS), Procedures, Guidance and Information (PGI), Section 234.201, "Policy," December 7, 2011.

U.S. Government Accountability Office, *DoD's Practices and Processes for Multiyear Procurement Should Be Improved*, Washington, D.C., GAO-08-298, 2008. As of June 21, 2017:
http://www.gao.gov/assets/280/272010.pdf

Utgoff, Kathleen P., and Dick Thaler, *The Economics of Multiyear Contracting*, Washington, D.C.: Center for Naval Analyses, 1982.

Wang, Chong, and Joseph San Miguel, "Unintended Consequences of Advocating Use of Fixed-Price Contracts in Defense Acquisition Practice," *Proceedings of the Eighth Annual Acquisition Research Symposium, Wednesday Sessions*, Vol. 1, Naval Postgraduate School, 2011.

Wilson, R. H., "A Scientific Routine for Stock Control," *Harvard Business Review*, Vol. 13, No. 1, 1934.

Younossi, Obaid, Mark V. Arena, Kevin Brancato, John C. Graser, Benjamin W. Goldsmith, Mark A. Lorell, Fred Timson and Jerry M. Sollinger, *F-22A Multiyear Procurement Program: An Assessment of Cost Savings*, Santa Monica, Calif.: RAND Corporation, MG-664-OSD, 2007. As of February 23, 2016:
http://www.rand.org/pubs/monographs/MG664.html